The Siddhāntasundara of Jñānarāja

The Siddhāntasundara of Jñānarāja

An English Translation with Commentary

TOKE LINDEGAARD KNUDSEN

Johns Hopkins University Press
Baltimore

Printed in the United States of America on acid-free paper
2 4 6 8 9 7 5 3 1

Johns Hopkins University Press
2715 North Charles Street
Baltimore, Maryland 21218-4363
www.press.jhu.edu

ISBN-13: 978-1-4214-1442-3 (hardcover)
ISBN-10: 1-4214-1442-2 (hardcover)
ISBN-13: 978-1-4214-1443-0 (ebook)
ISBN-10: 1-4214-1443-0 (ebook)

Library of Congress Library Control Number: 2013954357

With love and gratitude to Cecelia

Contents

List of Figures

List of Tables

Preface

My work with the *Siddhāntasundara* of Jñānarāja has been a journey of many years. Nearly a decade and a half ago, in 2000, I corresponded with David Pingree about my plans for combining my background in mathematics with my studies of the Sanskrit language to pursue a PhD degree on mathematics in India. Pingree suggested some possible topics for a dissertation, one of them being "the *Siddhāntasundara* of Jñānarāja, the last important siddhānta not to have been published."

I subsequently, in 2002, joined the doctoral program in the Department of History of Mathematics at Brown University, with David Pingree as my advisor. Soon after I enthusiastically embraced the suggestion to work on the *Siddhāntasundara*. I felt that not only would such a project nurture my interest in working with Sanskrit manuscripts, but also that working through a comprehensive Sanskrit treatise on astronomy would expose me to all the dimensions of the Indian astronomical tradition. Pingree made his copies of manuscripts of the *Siddhāntasundara* available to me, and I began the painstaking process of reading the old handwriting and making sense of the Sanskrit text.

Now, after more than a decade of study of Indian astronomy and the *Siddhāntasundara*, I am ready to present my English translation of and commentary on the *Siddhāntasundara*. The work on the *Siddhāntasundara* has been both tedious and demanding at times. I struggled with making sense of difficult Sanskrit passages and with adequately explaining an unfamiliar astronomical tradition to a reader. In many places I feel that great depth has been reached, but elsewhere my efforts still seem lacking. It is my sincere hope that my translation will inspire more work on the *Siddhāntasundara* and its influence, and that whatever errors I have made will be corrected by future generations of scholars.

The *Siddhāntasundara* is an important treatise, which stands at a unique point in time in the history of Indian astronomy, its composition marking the beginning of what has been called an early modern period in Indian astronomy. New ideas and perspectives are brought forth in it, and while maintaining tradition, it also breaks from it. As such, it feels as if the present translation only scratches the surface—both the background of the *Siddhāntasundara* and its influence in the following centuries deserve more study. It is my hope to be able to continue my work on this, especially by examining Jñānarāja's sources carefully.

The present book contains only an English translation of and commentary on the *Siddhāntasundara*, not the Sanskrit text. For a scholar it is naturally desirable to have access to the original source, and I therefore plan to publish a complete critical edition of the text of the *Siddhāntasundara* to accompany the translation.

During my journey with Jñānarāja and the *Siddhāntasundara*, which has led to the completion of this book, I have been aided by many. First and foremost, my thoughts go to David Pingree, my late mentor, who shared his vast erudition with me and taught me all that I know about Indian astronomy. I am sad that he did not see my dissertation completed and will not see this book in print, but I hope that he would have been pleased by my efforts. I would also like to thank Isabelle Pingree, who kindly opened up the Pingree home to me and made me feel welcome. I am also extremely thankful to Kim Plofker, friend and mentor, who took it on herself to oversee the completion of my dissertation after Pingree's death in 2005, and without whose support I would not have been able to make it. Kenneth Zysk has been generous and encouraging from the very beginning, and for that I will always be grateful. I am also grateful to Christopher Minkowski for his generous help and all our conversations about Jñānarāja, including our joint work on the *ṛtuvarṇana* of the *Siddhāntasundara*. Yann and Clemency Montelle and their three children have been true friends from graduate school to the present, and I am grateful for all that they have done for me. S. R. Sarma has been generous with his knowledge and time, and I have learned much from him. Keith Jones was always ready to help with technical issues when I was typesetting the manuscript for this book in LaTeX. I am grateful to my mother Gyda for all the loving support that she has given me throughout my life. At Johns Hopkins University Press, thanks are due to Trevor Lipscombe (now with the Catholic University of America Press) and Vincent Burke for all the support and encouragement that they have given me.

My wife Cecelia has been with me since the beginning of this book project and our daughter Ida was born in the midst of the writing process. Having a husband reading technical Sanskrit and a daughter building towers with Sanskrit-character letter blocks must have been quite an experience. I am ever thankful for all the love, patience, and understanding that Cecelia has shown me, and to her I dedicate this book.

The Siddhāntasundara of Jñānarāja

Introduction

Astronomy is a science that has been practiced in India since ancient times. It has served as a companion to astrology, providing the methods by which planetary configurations could be computed for a given time, or a lunar eclipse predicted. Astronomy was also practiced in the service of ritual and used to determine the cardinal directions and the correct timing for a sacrifice. Cosmology has been a close companion of astronomy in the Indian tradition, providing the theoretical model that serves as the basis for the mathematical formulae of astronomy.

Indian treatises on astronomy have come down to us since early times. There are different types of such treatises, one being called by the Sanskrit term *siddhānta*.[1] In classical Indian astronomy, a *siddhānta* is a comprehensive treatise that not only provides the reader with the mathematical formulae needed to compute, say, planetary positions and lunar eclipses, but also presents the underlying theory and model. Two well-known *siddhānta*s are Brahmagupta's *Brāhmasphuṭasiddhānta* from 628 CE and Bhāskara II's *Siddhāntaśiromaṇi* from 1150 CE.

Another *siddhānta*, less well known, is the *Siddhāntasundara*, "Treatise Beautiful," composed by the astronomer and scholar Jñānarāja around 1503 CE. The present book presents a translation of the *Siddhāntasundara* for the first time in English, accompanied by an introduction and commentary on the text.

I.1 Life and times of Jñānarāja

Not much is known about Jñānarāja, who wrote the *Siddhāntasundara*. The available information about Jñānarāja—when and

[1] Various types of astronomical treatises in India are on page 19.

where he lived, and his family background—is scant and scattered in various places, including the *Siddhāntasundara* itself and the writings of Jñānarāja's son Sūryadāsa.[2]

I.1.1 Brief overview of the *Siddhāntasundara*

Before listing the sources that have bearing on the date, location, and other information about Jñānarāja and the *Siddhāntasundara*, some brief notes on the *Siddhāntasundara* are given; an extended discussion of the structure of the text is given below (page 22).

The *Siddhāntasundara*, as composed by Jñānarāja, consists of three main parts, here termed "chapters," each of which are subdivided into a number of what we will here term "sections." The first chapter of the *Siddhāntasundara* is titled the *golādhyāya*, "Chapter on Cosmology"; the second the *grahagaṇitādhyāya*, or simply the *gaṇitādhyāya*, "Chapter on Mathematical Astronomy"; and the third the *bījagaṇitādhyāya*, "Chapter on Mathematics." However, the tradition has handed down the *golādhyāya* and the *gaṇitādhyāya* together as a single text, whereas the *bījagaṇitādhyāya* has been handed down as a separate text. The situation is similar to that of Bhāskara II, whose mathematical works, the *Līlāvatī* and the *Bījagaṇita*, form separate texts in the tradition, though it has been argued that they originally were part of Bhāskara II's astronomical masterpiece, the *Siddhāntaśiromaṇi*. Owing to the different nature of the *bījagaṇitādhyāya* and its treatment as a separate text in the Indian tradition, it is not included in the present translation, which covers the astronomical and cosmological parts of the *Siddhāntasundara*.[3]

References to passages in the *Siddhāntasundara*

Frequent references to passages in the *Siddhāntasundara* are given in the following. As is the case with all *siddhāntas*, the *Siddhāntasundara* is written in verse. Each section of the work contains a number of verses, numbered starting with 1. A verse can therefore be specified by the number of the chapter, the number of the section in which it is found, and its number in that sec-

[2]Information about Sūryadāsa is given on page 7.

[3]More details are given below (see page 22).

tion. As such "*Siddhāntasundara* 1.2.3" refers to the third verse of the second section in the first chapter (Chapter on Cosmology), and "*Siddhāntasundara* 2.1.17" refers to the 17th verse of the first section of the second chapter (Chapter on Mathematical Astronomy). The letters a, b, c, and d are used as superscripts to indicate the four quarters in verses of the *Siddhāntasundara*. For example, the reference $2.5.11^{\text{b}}$ indicates the second quarter of verse 11 of section 5 in the Chapter on Mathematical Astronomy, and similarly $2.5.11^{\text{c–d}}$ refers to the third and fourth quarters of the same verse.

I.1.2 Sources on Jñānarāja's date and location

The sources of information on Jñānarāja's date and location and about the *Siddhāntasundara* are as follows:

Source 1 Jñānarāja himself has four verses at the conclusion of the *bījagaṇitādhyāya* of the *Siddhāntasundara* which give a list of his ancestors and some information about his family.[4]

Source 2 The *Siddhāntasundara* contains an epoch, that is, a fixed point in time with reference to which future computations of planetary positions are made.[5]

Source 3 Jñānarāja's son Sūryadāsa provides information in several places. At the end of the *Sūryaprakāśa*, a commentary on the *Bījagaṇita*, a work on algebra by the mathematician and astronomer Bhāskara II, Sūryadāsa gives three verses, in which he gives information about the place where his family lived.[6]

---

[4]The text of the *bījagaṇitādhyāya* has not been published, but the four verses are quoted in Weber's description of the manuscript Berlin 833; see Weber 1853, 231–232.

[5]See *Siddhāntasundara* 2.2.58–64 and the accompanying commentary.

[6]No complete edition of the *Sūryaprakāśa* is available, but parts of it have been edited and published in Jain 2001 and Patte 2004. The three verses referred to are not included in the two partial editions, but they have been quoted in descriptions of manuscripts of the *Sūryaprakāśa*; see Eggeling 1886–1904, 1010, entry 2823 and Peterson 1892, Extract 529.

Source 4 In a verse in the *Prabodhasudhākara*, a work on the philosophical system known as *vedānta*, Sūryadāsa states from where his father hailed.[7]

Source 5 In two verses at the end of the *Gaṇitāmṛtakūpikā*, a commentary on the *Līlāvatī*, a mathematical work by Bhāskara II, Sūryadāsa gives information about the place where his father and grandfather lived.[8]

Source 6 In the *Paramārthaprapā*, a commentary on the *Bhagavadgītā*, a very important religious text in Hinduism, Sūryadāsa has a verse that talks about the place where Jñānarāja was born.[9]

Source 7 Sūryadāsa gives his own birth date in a verse at the end of the *Sūryaprakāśa*.[10]

The above sources will be referred to in the following.

I.1.3 The date of Jñānarāja

No date for the composition of the *Siddhāntasundara* or Jñānarāja's birth is given directly in the *Siddhāntasundara* or elsewhere. However, an epoch, that is, a fixed point in time with reference to which future computations of planetary positions are made, is given in the text (see Source 2 above). The epoch given is 1503 CE. Since epochs are used to simplify astronomical calculations and are easier to use the closer one is in time to the time defined by them, astronomical texts generally have epochs that roughly coincide with the date of their composition. We can therefore infer that the *Siddhāntasundara* was composed around that time. In other words, the *Siddhāntasundara* dates to around 1503 CE, or the beginning of the sixteenth century.

That Jñānarāja was an active scholar and composed the *Siddhāntasundara* at that time is consistent with the fact that

[7] The *Prabodhasudhākara* has not been published, but the verse in question is quoted from a manuscript by Sarma; see Sarma 1950, 224.

[8] The verses are cited in Eggeling 1886–1904, 1005, entry 2809.

[9] See Lallurama, Bakre, and Gokhale 2001, 3.1327.

[10] Only parts of the *Sūryaprakāśa* have so far been published (see Jain 2001 and Patte 2004), but the relevant verse is not therein. It has, however, been quoted in descriptions of the manuscripts of the *Sūryaprakāśa* given in Eggeling 1886–1904, 1010, entry 2823 and Peterson 1892, Extract 529.

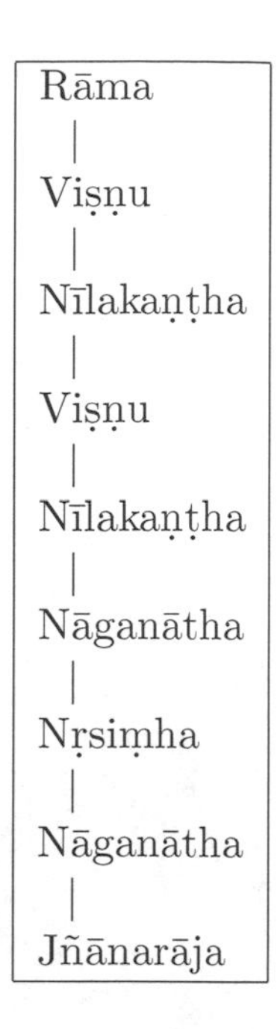

Table I.1: Jñānarāja's genealogy

Jñānarāja's son Sūryadāsa was born in 1507 or 1508 CE[11] according to his own testimony (see Source 7 above).

I.1.4 Jñānarāja's family

In the *bījagaṇitādhyāya* of the *Siddhāntasundara* (see Source 1 above), Jñānarāja gives a list of his male ancestors from his father Nāganātha to his great-great-great-great-great-great-grandfather Rāma, which is presented in Table I.1.

Jñānarāja's passage contains more information than names. He states that he belonged to a *brāhmaṇa* family of the Bhāradvāja *gotra*. A *brāhmaṇa* is a member of the highest social class in Hinduism, who traditionally is occupied with scholarship and priestly duties. A *brāhmaṇa* will trace his lineage back to a particular sage, which in the case of Jñānarāja's family is the sage

[11]More specifically, Sūryadāsa writes that he composed his commentary on Bhāskara II's *Bījagaṇita* in *śaka* 1460 when he was 31 years old. This means that he was born in either *śaka* 1429 or *śaka* 1430, that is, in 1507 or 1508 CE.

Bhāradvāja. In other words, the family belonged to the Bhāradvāja *gotra*.

Other than what Jñānarāja tells us (namely, their names and that they were learned men) we know nothing about his ancestors between Rāma, the most distant ancestor, and Jñānarāja's father Nāganātha, and no works composed by any of them are listed by Pingree.[12] However, a bit more can be said about Rāma and Nāganātha.

Jñānarāja's ancestor Rāma

Rāma, the most remote ancestor in Jñānarāja's genealogy, hailed from a town called Pārthivapura, on the northern bank of the Godāvarī River near a sacred place called Pūrṇatīrtha, and was honored by King Rāma, the ruler of Devagiri. Gaṇeśa, a member of Jñānarāja's family who composed two works on astrology, likewise connects his ancestors with King Rāma of Devagiri.[13] Aufrecht and Pingree identify this King Rāma as King Rāmacandra of the Yādava dynasty, who reigned 1271–1311 CE.[14] The nature of the honor bestowed on Jñānarāja's ancestor Rāma by King Rāmacandra of Devagiri is not specified, nor are the services that he performed in order to earn it, but it was possibly a land grant by which the family was able to support itself and its scholarship.

Jñānarāja's father Nāganātha

Jñānarāja describes his father Nāganātha as a learned man, who was fond of meditating on the smiling face of the god Gaṇeśa at the bank of the Godāvarī River. Sūryadāsa describes Nāganātha as accomplished in *jyotiḥśāstra* (the astral sciences) and in computation (see Sources 3 and 5 above). In other words, he was

[12]See Pingree 1981, 124, Table 9.

[13]Gaṇeśa flourished in the second half of the sixteenth century CE (see Pingree 1970–94, A.2.107–110, A.3.28, A.4.75–76, A.5.74–75, especially Pingree 1970–94, A.3.28, for Gaṇeśa's date). The verses in which Gaṇeśa gives information about himself and his family are the two last verses of the *Tājikabhūṣaṇa* and *Ratnāvalīpaddhati* 8.12–14, all of which are quoted in Pingree 1970–94, A.2.109.

[14]See Aufrecht 1962, 1.505, entry "Rāma of Pārthapura" and Pingree 1981, 120, n. 6. Note that there is only one king named Rāma or Rāmacandra in the genealogy of the Yādavas of Devagiri given in Verma 1970, 368.

likely an astronomer like his son Jñānarāja and his two grandsons. It is therefore possible that Jñānarāja's family was one of astronomers, but this is not stated explicitly by Jñānarāja or Sūryadāsa. No works composed by Nāganātha are known.

Jñānarāja's sons

In addition to these names of ancestors, it is known that Jñānarāja had at least two sons, Sūryadāsa, who has already been mentioned, and Cintāmaṇi. Both sons were learned and well versed in astronomy.

Sūryadāsa

Also known as Sūrya Paṇḍita, or simply Sūrya, Sūryadāsa was a polymath and a prolific writer on many subjects.[15] He is the most widely known member of Jñānarāja's family. His works include commentaries on the works of the mathematician and astronomer Bhāskara II, poetic works, a commentary on religious works sacred to Hinduism such as the *Bhagavadgītā* and the *veda*s, works on astronomy, and more.[16] He is credited with the invention of a genre of poetry called *vilomakāvya*, in which each verse can be read both from left to right and from right to left, giving two different narratives.[17] So famed was Sūryadāsa's learning that a poetical work entitled *Sūryodayakāvya* (literally, "The Rise of Sūrya"; the title is a play on the Sanskrit word *sūrya*, which in addition to being a proper name can mean "the sun") was written about his life and deeds.[18] As already noted, Sūryadāsa tells us that he was born in 1507 or 1508 CE.

Cintāmaṇi

The only known work of Jñānarāja's son Cintāmaṇi is a voluminous commentary on his father's *Siddhāntasundara* entitled

[15]For current information about Sūryadāsa, see Dikshit 1969–81, 2.144–145, Minkowski 2002, 507–508, Minkowski 2004a, 364–367, and Minkowski 2004b, 329–330.

[16]Regarding the works of Sūryadāsa, see Sarma 1950, 222–224.

[17]See Minkowski 2004b.

[18]The *Sūryodayakāvya* was published in Lalye 1979. The text has not yet been studied in detail.

Grahagaṇitacintāmaṇi.[19] In the *Grahagaṇitacintāmaṇi*, Cintāmaṇi elaborates on and defends his father's position on the *virodhaparihāra* issue (see page 40). He further writes amply on the properties of objects, and he describes physical experiments. Most important, however, is his attempt at integrating *jyotiḥśāstra* (the astral sciences), traditionally on the periphery of the *śāstras*, the "knowledge systems" of India, with the other *śāstras*. In particular, he recasts many of Jñānarāja's arguments and demonstrations in the language of the philosophical *śāstras*.[20] Beyond his authorship of the *Grahagaṇitacintāmaṇi*, nothing is known of Cintāmaṇi.[21]

Jñānarāja's wife

In the *Sūryodayakāvya*, the name of Jñānarāja's wife is given as Ambikā,[22] though no further details are given.

I.1.5 Pārthapura, Jñānarāja's native town

The place where Jñānarāja's family lived has been identified by Jñanaraja and Suryadasa both by name and by proximity to other locations. According to a passage from the *bīja-gaṇitādhyāya* (see Source 1 above), Jñānarāja's ancestor Rāma resided in a town called Pārthivapura. In the verses quoted from the works of Sūryadāsa (see Sources 2–3), Nāganātha, Jñānarāja's father, is said to have lived in Pārthapura, and presumably it is understood that Jñānarāja did as well. There is no doubt that Pārthivapura and Pārthapura are the same place, which we will refer to as Pārthapura in the following. In the *Prabodhasudhākara* (see Source 4 above), Sūryadāsa says that Jñānarāja lived in a village called Prastara, but this is most likely the village outside the town of Pārthapura where Jñānarāja lived.

Jñānarāja's town is mentioned in other sources as well. It is called Pārthavālapura in the *Sūryodayakāvya*,[23] Pātharī by

[19] See Pingree 1970–94, A.3.49, A.4.94, Pingree 1981, 30, and Minkowski 2004a, 360. The *Grahagaṇitacintāmaṇi* has not been edited or fully studied.

[20] See Minkowski 2002, 504–506 and Minkowski 2004a, 361–362.

[21] For current information about Cintāmaṇi, see Pingree 1970–94, A.3.49, A.4.94, Minkowski 2002, 504–507, and Minkowski 2004a, 360–364.

[22] See *Sūryodayakāvya* 1.18 in Lalye 1979.

[23] *Sūryodayakāvya* 1.15.

the astronomer Kamalākara (fl. about 1658 CE),[24] and Pāthrī in Islamic sources. It is clear that these are just variant names of Pārthapura.

Later on, the family relocated from Pārthapura to Bid, a nearby town.[25]

Location of Pārthapura

About the location of Pārthapura we are told the following by Jñānarāja and Sūryadāsa: Pārthapura is on the northern bank of the Godāvarī River (the modern Godavari River in south-central India) near a *tīrtha* (a sacred place) called Pūrṇatīrtha. More specifically, Sūryadāsa writes that Pārthapura is located one *krośa* (about $2\frac{1}{4}$ miles, or a bit more than $3\frac{1}{2}$ kilometers) to the west of the confluence of the Maṅgalā and the Godāvarī Rivers on the northern bank of the Godāvarī.

Pūrṇatīrtha is extolled in the *Gautamīmāhātmya*, a section of the *Brahmapurāṇa* praising the Godāvarī River (also known as the Gautamī River).[26] According to this text, Pūrṇatīrtha is on the northern bank of the Godāvarī, a dwelling place of the deities Viṣṇu and Śiva, and at the confluence of the Maṅgalā and the Godāvarī Rivers.[27]

There has been some confusion among scholars about the rivers mentioned by Sūryadāsa in the sources pertaining to the history of his family, and it is appropriate to address that here. When Sūryadāsa mentions the Gaṅgā, the Sanskrit name for the Ganges, he is not referring to the Ganges in north India. He is instead referring to the Godāvarī, which is considered to have the same source as the Ganges and is therefore often referred to simply as Gaṅgā in texts from the regions of the Deccan Plateau. Furthermore, Sūryadāsa makes mentions of Vidarbhā and Maṅgalāgaṅgā. Dikshit believes that the Vidarbhā

[24]See Dikshit 1969–81, 2.151–152. For Kamalākara, see Pingree 1970–94, A.2.21–23, A.3.18, A.4.33, A.5.22.

[25]See Dikshit 1969–81, 2.143. According to the *Sūryodayakāvya*, Sūryadāsa went to Bid, so probably it was with him that the family (or part of it) moved there (see Lalye 1979, iii).

[26]*Brahmapurāṇa* 122 (*Gautamīmāhātmya* 53). The edition used is Schreiner and Söhnen 1987.

[27]*Brahmapurāṇa* 122.1–2 and 122.100 (*Gautamīmāhātmya* 53.1–2 and 53.100).

is the same river as the Maṅgalā,[28] and in the *Catalogue of the Sanskrit Manuscripts in the Library of the India Office*, the Maṅgalāgaṅgā is taken as indicating one river and identified with the Vidarbhā.[29] However, the word *maṅgalāgaṅgā* is not to be read as indicating just one river, but rather as indicating two, as Gaṅgā here refers to the Godāvarī River. Furthermore, the *Brahmapurāṇa* clearly considers the Vidarbhā and the Maṅgalā to be two separate rivers. Whereas the Maṅgalā River is mentioned in connection with Pūrṇatīrtha, the confluence of the Vidarbhā and the Godāvarī is praised in a separate chapter.[30] It is clear, though, that the two rivers are not far from each other.

Identification of Pārthapura

Dikshit identifies Pārthapura as the modern Pathri, which Pingree places in the Parbhani District of Maharashtra.[31] Lalye similarly identifies Pārthavālapura with Pathri.[32] The identification seems certain.

History of Pārthapura

Pārthapura is the ancestral home not only of Jñānarāja's family, but also of the Islamic Nizām Shāhī dynasty that ruled the Ahmadnagar kingdom. The following account of the history of Pārthapura is drawn from Islamic sources, and since these refer to Pārthapura as Pāthrī, we will use that name in the following.

According to the Muslim historian Firishta (fl. about 1560–1620 CE), Ahmad Nizām Shāh (r. 1490–1509),[33] the founder of the Nizām Shāhī dynasty, was the son of a *brāhmaṇa* named Tīma (or Timapa), who received the name Malik Hasan Bahrī when he converted to Islam. He was the son of the *brāhmaṇa* Bhairava, who worked as a Kulkarni, that is, a hereditary village accountant, in Pāthrī.[34]

[28]See Dikshit 1969–81, 2.141.

[29]See Eggeling 1886–1904, 1004, entry 2809 and Eggeling 1886–1904, 1010, entry 2823.

[30]*Brahmapurāṇa* 121 (*Gautamīmāhātmya* 52).

[31]For Dikshit's identification, see Dikshit 1969–81, 2.141. For Pingree, see Pingree 1981, 120.

[32]See Lalye 1979, iii.

[33]See Haig 1928, 704.

[34]See Firishtah 1971, 3.116, 130, Haig 1928, 398, and Benson 2001, 108.

Ahmad Nizām Shāh was succeeded by his son Burhān Nizām Shāh (r. 1509–1553),[35] who was approached by his relatives in Pāthrī who "wished to enjoy the protection and patronage of their royal kinsman."[36] At that time, Pāthrī was on the frontier between the Islamic kingdoms of Ahmadnagar and Berar, but belonged to the latter. In keeping with his relatives' desire, Burhān Nizām Shāh offered 'Alā-ud-dīn 'Imād Shāh, the king of Berar, a favorable exchange for the town of Pāthrī. The request was denied, however, and 'Alā-ud-dīn 'Imād Shāh began fortifying the town. Burhān Nizām Shāh subsequently, in 1518, invaded and conquered Pāthrī, which he held until 1527, when 'Alā-ud-dīn 'Imād Shāh retook the city.[37] The army of Berar was, however, unable to hold it, and in that same year it fell again to Burhān Nizām Shāh. Firishta writes that Burhān Nizām Shāh destroyed the fort of Pāthrī after a siege lasting two months, and proceeded to give the district in charity to his *brāhmaṇa* relatives there. Burhān Nizām Shāh's relatives kept the town until the time of the Mughal Emperor Akbar (1542–1605).[38] 'Alā-ud-dīn 'Imād Shāh appealed for help to the king of Gujarat,[39] whose intervention forced Burhān Nizām Shāh to agree to a peace agreement, according to which he was, among other things, to return Pāthrī to Berar. This, however, never happened.[40] In 1596, a Mughal army defeated the army of Ahmadnagar at Pāthrī.[41]

These events took place after the *Siddhāntasundara* was written, but during the times when Jñānarāja's sons, Cintāmaṇi and Sūryadāsa, were active scholars. The events illustrate the situation in the area at the time, including the time of Jñānarāja, which saw general unrest between rivaling Islamic kingdoms. However, following the tradition of *brāhmaṇa* writings, Jñānarāja makes no mention at all of the sociopolitical environment of his times.

For the Kulkarni title, see Firishtah 1971, 3.130, n. 93.

[35]See Haig 1928, 704.

[36]See Haig 1928, 435. According to Firishta,they desired to have their ancestral rights (that is, the position as hereditary village account) restored (see Firishtah 1971, 3.130).

[37]See Haig 1928, 435.

[38]See Firishtah 1971, 132.

[39]See Haig 1928, 324.

[40]See Haig 1928, 325, 436.

[41]See Pingree 1997, 84.

I.1.6 The works of Jñānarāja

When examining manuscript catalogues or other modern works, one generally sees two works attributed to Jñānarāja, namely, the *Siddhāntasundara* and a work on mathematics referred to as the *Bījādhyāya*,[42] or the *Sundarasiddhāntabīja*.[43] The relationship between these two texts will be discussed below.[44]

In addition, Sūryadāsa has a verse in the *Sūryaprakāśa* that describes his father's literary output (see Source 3 above). In the verse, Sūryadāsa credits his father with writing four works, namely, an astronomical treatise entitled *Siddhāntasundara*, as well as three unnamed works on horoscopy (*jātaka*), rhetoric (*sāhitya*), and the art of singing (*gītaśāstra*). Since the *Siddhāntasundara* is mentioned first and no titles are given for the other three works, it is reasonable to assume that Sūryadāsa considered the *Siddhāntasundara* to be Jñānarāja's *magnum opus*.

Besides Sūryadāsa's testimony, there is no information on or further references to any works on rhetoric and song authored by Jñānarāja. However, the *Catalogus Catalogorum* attributes a work entitled *Yavanajātaka* to Jñānarāja,[45] an attribution that is repeated in the *New Catalogus Catalogorum*.[46] Unfortunately, the source given in the *New Catalogus Catalogorum* is a verse from the *Vṛddhayavanajataka* of Mınarāja which has wrongly been attributed to Jñānarāja.[47] In other words, the passage quoted in the *Prauḍhamanoramā* is not from a work by Jñānarāja.

---

[42]See, for example, Pingree 1981, 64. In Pingree 1970–94, A.3.75, the work is referred to as "a *Bījādhyāya* for the *Siddhāntasundara*."

[43]See Sen 1966, 94.

[44]See page 22.

[45]Aufrecht 1962, 2.43.

[46]The latter notes that the *Yavanajātaka* of Jñānarāja is quoted in the *Prauḍhamanoramā* of Divākara, a commentary on the *Jātakapaddhati*, an astrological work composed by Keśava (fl. 1496/1507 CE). Keśava flourished in Nandigrāma, the modern Nandod in Gujarat (see Pingree 1970–94, A.2.65–74, A.3.24, A.4.64–66, A.5.56–59 and Pingree 1973), while Divākara was born in 1606 CE and wrote the *Prauḍhamanoramā* in 1626 (see Pingree 1970–94, A.3.106–110, A.4.111, A.5.141–142 and Pingree 1981, 125, Table 11).

[47]The passage referred to in the *New Catalogus Catalogorum* as a reference to a *Yavanajātaka* of Jñānarāja must be the *Prauḍhamanoramā* on *Jātakapaddhati* 33 (see Vámanáchárya 1882, 126), but the passage cited is actually *Vṛddhayavanajātaka* 7.61 (see Pingree 1976b, 1.54).

I.2 The *Siddhāntasundara*

As noted above, the *Siddhāntasundara*, Jñānarāja's only known work, is an extensive astronomical and cosmological treatise. The treatise is also referred to as the *Sundarasiddhānta* in the manuscripts, but in the following we shall refer to it exclusively as the *Siddhāntasundara*.

Before proceeding to a discussion of the text and its contents, a brief overview of Indian astronomy will be given. The overview introduces key terms, so it can be used as a reference when reading the *Siddhāntasundara*; the material will be treated in greater detail in the commentary to the text.

I.2.1 Indian astronomy

In the tradition of India, astronomy is part of one of the traditional *śāstra*s (knowledge systems), namely, *jyotiḥśāstra*. In addition to mathematics and astronomy, *jyotiḥśāstra* includes various divinatory arts, including astrology. Jñānarāja discusses the divisions of *jyotiḥśāstra* in the *Siddhāntasundara*.[48]

The model behind Indian astronomy is geocentric; a spherical earth is at the center of a spherical universe, bounded outwardly on the sphere of the stars, that is, the sphere on which the fixed stars are found. In between the earth and the sphere of the stars the planets orbit the earth. Note that in accordance with ancient terminology, here and in the following the word "planet" will denote either one of the two luminaries, that is, the sun and the moon, or one of the five planets known to the ancient Indians, namely, Venus, Mercury, Mars, Jupiter, and Saturn; if it is necessary to distinguish further, the former are called "luminaries" and the latter "star-planets."

When giving the location of a planet, Indian astronomy generally gives a longitude with respect to the ecliptic. It is important to note in this regard that Indian astronomy operates with a sidereal zodiac. This means that the position of the sun and the other planets is measured with respect to the fixed stars, not, as in the tropical zodiac of Greek astronomy, with respect to the vernal equinox.

---

[48] See *Siddhāntasundara* 1.1.8.

Epicyclic model

The Indian astronomical system seeks to describe planetary motion, including predicting planetary positions at a given time, to compute lunar and solar eclipses, and so on. The model employed is geocentric, that is, the center of a spherical earth is viewed as the center of the universe, and uses epicycles to describe the motion of the planets.

An epicycle is a circle, the center of which moves on another circle that has its center at the center of the earth. In the case of the two luminaries, the planet moves on the epicycle, whereas in the case of the star-planets, two epicycles are used to describe the planet's motion (in modern terms we can say that one epicycle is accounting for the fact that the planet orbits the sun, not the earth, and the second for the fact that the orbit is not circular but elliptical). The model is a pre-Ptolemaic Hellenistic model following certain Aristotelian principles which was transmitted to India in the first half of the first millennium CE, but it also contains certain distinct Indian features, such as the motion of the planets being caused by cosmic winds.[49]

In Indian astronomy, each of the star-planets has two epicycles to describe their motion; one is called *manda* ("slow"), the other *śīghra* ("fast"). The two luminaries have only one epicycle, the *manda*.

As a consequence of the model, a planet changes it distance to the earth over time. When the planet is as far as it can get from the earth, it is said to be at its **apogee**; similarly, it is at its **perigee** when it is the closest it can get to the earth.

A more detailed description of the planetary model of Indian astronomy can be found in the commentary to *Siddhāntasundara* 2.2.15.

Planetary motion

The sun's orbit around the earth is called the **ecliptic**. It differs from the **celestial equator** (the terrestrial equator projected out to meet the sphere of the stars). More specifically, both the ecliptic and the celestial equator are great circles on the sphere of the stars, but the ecliptic is tilted from the celestial equator

[49] See *Siddhāntasundara* 1.1.64, where the winds are introduced by Jñanarāja.

by an angle known as the **obliquity of the ecliptic** ε, which in the Indian system is $\varepsilon = 24°$. The ecliptic and the celestial equator intersect at two points: the **vernal equinox**, where the sun crosses the celestial equator from south to north, and the **autumnal equinox**, where it crosses from north to south.

As the sun travels across the ecliptic, its distance to the celestial equator varies. The distance is measured as the angle from the position of the sun to the celestial equator and is called the **declination**. At the equinoxes, the declination is 0°, and so the sun is on the celestial equator. At the two points on the ecliptic that are 90° from an equinox, the sun is as far from the celestial equator as it can get; the point that is north of the celestial equator is the **summer solstice** and the point south of the celestial equator is the **winter solstice**. The declination of the summer solstice is $\varepsilon = 24°$ north, and similarly it is $\varepsilon = 24°$ south at the winter solstice.

Similarly, each planet has its own orbit around the earth. These orbits are close to the ecliptic but differ from it, being displaced by an angle unique to the particular planet. The two points where the orbit of the planet intersects the ecliptic are called the planet's **nodes**. The **ascending node** is the node where the planet crosses the ecliptic from south to north, and the **descending node** is where the planet crosses the ecliptic from north to south. The nodes of the moon are of importance when computing lunar eclipses.

A planet's **latitude** is the angle between the position of the planet and the ecliptic; the latitude can be either north or south. Like the sun, a planet has a declination, which is the angle from the position of the planet on the ecliptic to the celestial equator. The latitude must be taken into account also to get the actual distance between the celestial equator and the planet.

In addition to its own independent motion across the fixed stars, a planet is carried along with the rotation of the fixed stars. During the span of a day (a 24-hour period), a planet only moves slightly. However, during that time the planet is carried across the sky by the rotation of the fixed stars. In that process, it describes a circle, called its **diurnal circle**. The diurnal circle is generally not a great circle, but smaller.

Astronomical circles

For an observer of the sky at a given location of the earth, the **horizon** is the great circle that separates the half of the sky that he can see from the half that is hidden beneath the earth, outside of his view. The **zenith** of the observer is the point of the sky directly above him. The **meridian** for the observer's location is the great circle that passes through the North and the South Poles and the zenith. The **six o'clock circle** is the great circle perpendicular to the celestial equator and passing through the east and west points. The **prime vertical** is the great circle passing through the zenith and the east and west points. The celestial equator and the ecliptic have already been defined above.

Cosmic time periods

Indian astronomy operates with very long time periods. The Sanskrit word *yuga*, meaning "world age," is used in the designation of most of them. A ***kalpa*** is a period of 4,320,000,000 years. It consists of 1,000 ***mahāyugas*** of 4,320,000 years each. A *mahāyuga* is divided into a ***kṛtayuga*** of 1,728,000 years, a ***tretāyuga*** of 1,296,000 years, a ***dvāparayuga*** of 864,000 years, and a ***kaliyuga*** of 432,000 years. According to the Indian tradition, we currently live in a *kaliyuga*, the worst of these four ages.

The time periods of Indian astronomy are discussed in great detail in *Siddhāntasundara* 2.1.2–3 and the accompanying commentary.

The Indian calendar

Having a calendar is essential to the practice of astronomy. The Indian calendar is a complicated lunisolar calendar; that is, it takes into account both the phase of the moon and the year as defined by the motion of the sun. It operates with lunar months, that is, a month is the period from one conjunction of the sun and the moon to the next. During a lunar month, the moon moves farther from the sun until catching up with it again. When the moon has moved 12° away from the sun, a time period known as a ***tithi*** has passed. When the moon has moved another 12°

away from the sun (that is, when the angular distance between the two luminaries is 24°), another *tithi* has passed, and so on. There are therefore 30 *tithi*s in a lunar month. Note that since the sun and the moon change their speed, a *tithi* varies in length from about 19 hours to 26 hours.

The lunar month is divided into two parts called ***pakṣa*s**. The **bright *pakṣa***, with which the month begins, is the time from the conjunction of the sun and the moon (where the moon is not visible) to the opposition of the sun and the moon, that is, the following full moon. The **dark *pakṣa*** is the time from the full moon to the next conjunction of the sun and the moon.

A **civil day** is the period from one sunrise to the next (or from one midnight to the next according to some schools of Indian astronomy). It is divided into 60 ***ghaṭikā*s**, and each *ghaṭikā*, in turn, is divided into 60 ***pala*s**.

The *yojana*

A *yojana* is a measure of distance used in the Indian tradition. The value of a *yojana* differs in various contexts but is generally considered to be between $2\frac{1}{2}$ miles and 9 miles. It is used frequently in the *Siddhāntasundara*.

Schools of Indian astronomy

Indian astronomical texts are generally divided according to a number of *pakṣa*s, or schools. The various *pakṣa*s differ in certain details, including some astronomical parameters, such as the length of the year, and on details such as whether the day begins at sunrise or at midnight. However, beyond minor differences, the schools all agree on the basic astronomical model, as outlined briefly above.

There are four *pakṣa*s:

1. The *brāhmapakṣa*,[50] the oldest of the *pakṣa*s, is believed to be a revelation of the deity Brahmā. Its founding text is a treatise entitled the *Paitāmahasiddhānta*,[51] which survives only in corrupted form in a larger compilation, the *Viṣṇudharmottarapurāṇa*. Two important texts of the *brāhma-*

[50]For the *brāhmapakṣa*, see Pingree 1978a, 555–589.
[51]See Pingree 1970–94, A.4.259–260.

pakṣa are Brahmagupta's *Brāhmasphuṭasiddhānta*[52] and Bhāskara II's *Siddhāntaśiromaṇi*.

2. The *āryapakṣa*[53] was founded by Āryabhaṭa, one of the most famous astronomers in Indian history, and its main text is his *Āryabhaṭīya*.[54] Like the *brāhmapakṣa*, Āryabhaṭa credits Brahmā with a revelation of astronomy.

3. The *ārdharātrikapakṣa*[55] is like the *āryapakṣa* founded by Āryabhaṭa, but its founding text is now lost. The name of the school is derived from its use of a midnight (*ardharātra*) epoch.

4. The *saurapakṣa*[56] is based on a *Sūryasiddhānta* from the late eight or early ninth century CE.

Jñānarāja's *Siddhāntasundara* belongs to the *saurapakṣa*.

Bhāskara II and his legacy

One of the most renowned figures in Indian astronomy and mathematics is Bhāskara II (b. 1114 CE).[57] A number of important astronomical and mathematical works were composed by him, and, as we saw above, descendants of his, who held high positions at the Yādava court, were instrumental in promoting the study of these works, and of astronomy in general.

The most notable astronomical work of Bhāskara II is the monumental *Siddhāntaśiromaṇi*,[58] which was completed in 1150 CE. As a treatise, it is clear and comprehensive, a testimony to Bhāskara II's deep understanding of the subject.

As noted by Pingree, the most impressive quality of the *Siddhāntaśiromaṇi* is its comprehensiveness,[59] a quality that en-

[52]For Brahmagupta and the *Brāhmasphuṭasiddhānta*, see Pingree 1970–94, A.4.254–257, A.5.239–240.

[53]For the *āryapakṣa*, see Pingree 1978a, 590–602.

[54]For Āryabhaṭa, see Pingree 1970, Pingree 1970–94, A.1.50–53, A.2.15, A.3.16, A.4.27–28, A.5.16–17, and Plofker and Knudsen 2008a.

[55]For the *ārdharātrikapakṣa*, see Pingree 1978a, 602–608.

[56]For the *saurapakṣa*, see Pingree 1978a, 608–618.

[57]For Bhāskara II and his works, see Pingree 1970–94, A.4.299–326, A.5.254–263. He is called the second Bhāskara to avoid confusion with an earlier astronomer of the same name.

[58]The edition of the *Siddhāntaśiromaṇi* used here is Śastrī 1999.

[59]See Pingree 1981, 26.

sured that it (and other of Bhāskara II's works) became normative in the period following him. In fact, after the composition of the *Siddhāntaśiromaṇi* in 1150 CE, the Indian astronomers moved in a didactic direction, focusing on writing commentaries on existing treatises and composing astronomical tables meant to facilitate the computation of planetary positions for casters of horoscopes and makers of calendars[60] rather than writing original treatises.[61]

A concept central to the work of Bhāskara II is that of *vāsanā*, "demonstration." A *vāsanā* for an astronomical algorithm or formula is a demonstration of it, that is, what we can loosely call a proof.[62] Bhāskara II's own commentary on the *Siddhāntaśiromaṇi* is called the *Vāsanābhāṣya*, "Demonstration Commentary," and it provides demonstrations for the mathematical algorithms and formulae found in the main text.

Types of astronomical treatises

As classified by the tradition, different types of astronomical treatises were composed in ancient and medieval India.

The most extensive type of Indian astronomical treatise is the so-called *siddhānta*. A *siddhānta* generally contains theory as well as mathematical algorithms and formulae. It bases its computations on the beginning of an enormous time period of 4,320,000,000 years known as a *kalpa* ("age").

Comparable to a *siddhānta*, a *tantra* also contains theory, but it bases its computations on the beginning of a smaller time period, the *kaliyuga*, which forms the last portion of a *kalpa* (according to the Indian tradition, we are currently in a *kaliyuga*). The distinction between a *siddhānta* and a *tantra* is not always clearly defined. For example, Jñānarāja refers to his *Siddhāntasundara*, a *siddhānta*, as a *tantra* numerous times.[63] Note that Jñānarāja discussed the difference between a *siddhānta* and a *tantra* in *Siddhāntasundara* 1.1.8, the distinction there being one between works authored by divine figures and works authored by human beings.

[60]See Pingree 1978a, 41.

[61]See Pingree 1981, 26.

[62]It should be noted, though, that the Indian tradition does not operate with proofs in the Euclidean sense that we use today.

[63]See, e.g., *Siddhāntasundara* 1.1.79.

A *karaṇa* is a smaller astronomical work. It does not contain theory the way that a *siddhānta* does, and its formulae are geared toward ease of practical application. The computations are based on an epoch close in time to the date of the composition of the *karaṇa*.

A *koṣṭhaka* is an astronomical table meant to facilitate astronomical computations for casters of horoscopes and makers of calendars.[64]

The cosmology of the *purāṇas*

As mentioned above, the cosmology of the Indian astronomical tradition features a spherical earth in the center of the universe. There were, however, other cosmologies in the broader Indian tradition. One of them is found in a class of sacred texts in Hinduism, the *purāṇas* (literally, "Old Stories").

The content of the *purāṇas* is encyclopedic, encompassing mythology, religion, divination, cosmology, ancient legends, and much more. According to Pingree, the source of the cosmological material in the *purāṇas* probably dates from the early centuries CE.[65] Owing to the sacred nature of these works, their cosmology was generally accepted in the larger Hindu world and thus exerted influence on the Indian astronomical tradition. Many of the concepts and much of the terminology of the cosmology of the *purāṇas* were incorporated into the cosmological model of the Indian astronomers. For this reason it is necessary to briefly outline the cosmology of the *purāṇas*.

According to the *purāṇas*, the earth is a flat disk at the center of which is a huge mountain called Meru. A circular region called Jambūdvīpa surrounds Meru, and at the southern part of Jambūdvīpa is Bhāratavarṣa, that is, India. Jambūdvīpa is surrounded by an annular ocean of salt water, and this ocean is, in turn, surrounded by alternating annular regions and oceans. There are seven annular oceans in total.

The planets revolve above the earth around an axis through the mountain Meru. In this model, the sun is closer to the earth than the moon, which is the opposite of what is taught in the astronomical system of ancient and medieval India. When the

---

[64]See Pingree 1978a, 41.

[65]See Pingree 1978a, 554–555.

sun is behind Meru, its light is blocked and it is night; when it is in front of Meru, it is day.

As is clear, the cosmological model of the *purāṇas* is very different from the model used by the Indian astronomers, and, in fact, the two cosmologies are inconsistent with each other. To give an example, if, as in the *purāṇas*, the sun is closer to us than the moon, it is no longer possible to explain solar eclipses as the disk of the moon covering the disk of the sun, as is done by the Indian astronomical tradition.

Traditionally, the approach of the Indian astronomers to the problem of the two cosmologies contradicting each other was to incorporate certain elements of the cosmology of the *purāṇas* into their model while rejecting others. As we shall see below regarding the issue of *virodhaparihāra* (page 40), Jñānarāja's approach was to seek a synthesis between the cosmology of the *purāṇas* and the cosmology of the Indian astronomical tradition which preserves the authority of the *purāṇas* (and therefore does not reject elements from them) while still maintaining the basic cosmological model in order to have a workable science.

Representation of numbers in Indian mathematics and astronomy

Since the *Siddhāntasundara* contains numerous formulae from mathematical astronomy, it is necessary to also briefly discuss how numbers are represented in Sanskrit treatises on mathematics and astronomy.

In Sanskrit astronomical texts, which are written in versified form, numbers are generally expressed using a system known as *bhūtasaṅkhyā* (literally, "object-numeral").[66] The *bhūtasaṅkhyā* system assigns a numerical value to certain Sanskrit words. For example, the Sanskrit word *danta* (tooth) represents the number 32 (after the thirty-two permanent adult teeth of a human being), the Sanskrit word *sāgara* (ocean) represents the number 4 (after the four oceans in the Indian tradition), and the Sanskrit word *śara* (arrow) represents the number 5 (after the five arrows of Kāmadeva, the Indian god of love).[67] Larger numbers can be

[66] See Sarma 2003, 38–41. While there are other systems, they are not used by Jñānarāja and will therefore not be discussed here.

[67] For a list of Sanskrit words corresponding to a given number, see Sarma 2003, 59–69.

created by combining two or more such words in a compound, which is read numerically from right to left. Using hyphens to separate the members of each compound for the sake of clarity, we have that *śara-danta* means 325, *danta-sāgara-śara-śara-śara* means 555,432, and *danta-śara-danta-danta-sāgara-śara* means 543,232,532.

Numbers expressed via the *bhūtasaṅkhyā* system are translated as numerals in the present translation. In other words, rather than translating the Sanskrit phrase *danta-sāgara-śara-śara-śara* as "teeth-oceans-arrows-arrows-arrows," it is given as its numeric equivalent 555,432.

Manuscripts of Sanskrit astronomical texts sometimes have numerals added to clarify the numbers expressed via the *bhūtasaṅkhyā* system.

Integral numbers are expressed using the same decimal system that we use today. Fractional parts of numbers, however, are not expressed in the decimal system, but rather in the sexagesimal system, that is, by using base 60 instead of base 10. In other words, in the Indian tradition, a number a is represented as

$$a = n + \frac{m_1}{60} + \frac{m_2}{60^2} + \ldots + \frac{m_k}{60^k}, \qquad \text{(I.1)}$$

where n, m_1, m_2, $\ldots$, and m_k are nonnegative integers, and $0 \leq m_i < 60$ where i is one of 1, 2 , $\ldots$, or k.

In modern texts, a number in an Indian text is often expressed as

$$a = n; m_1, m_2, \ldots, m_k. \qquad \text{(I.2)}$$

A semicolon is used to separate the integer parts of the number from the fractional part, while commas are used to separate each place of the fractional part in base 60. To give a specific example, the number $82\frac{23}{50} = 82.46$ is represented as 82;27,36, that is, $82 + \frac{27}{60} + \frac{36}{60^2}$, which is equal to 82.46. In the translation of the *Siddhāntasundara*, this notation will occasionally be used.

I.2.2 Structure of the *Siddhāntasundara*

The *Siddhāntasundara* has three main divisions, which we will refer to as "chapters," each subdivided into a number of what we will refer to as "sections." The three chapters of the *Siddhāntasundara* are as follows:

1. The *golādhyāya*—the Chapter on Cosmology;

2. The *grahagaṇitādhyāya*, or simply the *gaṇitādhyāya*—the Chapter on Mathematical Astronomy; and

3. The *bījagaṇitādhyāya*—the Chapter on Mathematics.

The *bījagaṇitādhyāya*

As has already been noted, in manuscript catalogues or other modern works, two works are generally attributed to Jñānarāja:

1. the *Siddhāntasundara*; and

2. a work of mathematics called the *Bījādhyāya*.

There is a somewhat ambiguous relationship between the *Siddhāntasundara* and this mathematical work, and it is not clear from the modern sources whether the mathematical work is a part of the *Siddhāntasundara*, or whether it is an independent work. Taking a cursory glance at the manuscript evidence, it is not difficult to see why the relationship between the *Siddhāntasundara* and the mathematical work is not entirely clear, for the scribal tradition, which preserved and copied the two works, has consistently handed them down entirely separately; there appears to be no single manuscript that contains them both.

It will be argued here that the *Bījādhyāya* was not intended to be an independent work by Jñānarāja, but in fact a part of the *Siddhāntasundara* entitled *bījagaṇitādhyāya*. However, the scribal tradition separated it from the astronomical and cosmological parts of the *Siddhāntasundara*, and it was passed down as a separate text. Similarly, the two mathematical works of Bhāskara II, the *Līlāvatī* and the *Bījagaṇita*, are sometimes considered to form part of his astronomical *magnum opus*, the *Siddhāntaśiromaṇi*.

No in-depth scholarly study of the *bījagaṇitādhyāya* has been undertaken, but some arguments that the text is a part of the *Siddhāntasundara* are presented here:

1. According to Dikshit, there is a statement in the *Grahagaṇitacintāmaṇi*, the commentary on the *Siddhāntasundara* by Jñānarāja's son Cintāmaṇi, that the *Siddhāntasundara*

contains the *bījagaṇitādhyāya*,[68] and Sūryadāsa makes a similar statement in the *Sūryaprakāśa*.[69]

2. The concluding verses of the *bījagaṇitādhyāya*'s sections have the same format as those for the *Siddhāntasundara*. They do not identify the sections as belonging to a *bījagaṇitādhyāya*, though.

3. The chapter colophons of the *bījagaṇitādhyāya* state that the text is a part of the *Siddhāntasundara*.

4. The final verses of the *bījagaṇitādhyāya* state directly that the title of the work is the *Siddhāntasundara*, and these verses are furthermore clearly the concluding verses of the work as a whole, that is, the concluding verses of the *Siddhāntasundara*.

As is clear from the above observations, the *bījagaṇitādhyāya* forms a part of the *Siddhāntasundara*, constituting the final part of that work.

However, since the *bījagaṇitādhyāya*, which is a work on pure mathematics, differs in topic and scope from the remainder of the *Siddhāntasundara*, and furthermore has been treated as a separate text by the scribal tradition of India, it has not been included in the present translation, which focuses on the astronomy and cosmology of Jñānarāja.

In the following, the title *Siddhāntasundara* will refer to only the *golādhyāya* and the *grahagaṇitādhyāya*, whereas the mathematical chapter will be referred to simply as the *bījagaṇitādhyāya*.

The *golādhyāya* and the *gaṇitādhyāya*

Not all of the manuscripts of the *Siddhāntasundara* used for the present translation include both the *golādhyāya* and the *gaṇitādhyāya*; some include only the former, others only the latter. Among the manuscripts that contain both, some begin with the *grahagaṇitādhyāya*, others with the *golādhyāya*. In other words, there is no consensus in the manuscript tradition as to the order of the two parts.

---

[68] See Dikshit 1969–81, 2.142.

[69] See Jain 2001, 45 of Sanskrit text.

That said, it is not difficult to deduce the proper order of the two chapters. In fact, the *golādhyāya* is the first chapter and the *grahagaṇitādhyāya* the second. The reasons for this order are as follows:

1. At the beginning of his commentary on the first section of the *grahagaṇitādhyāya*, Cintāmaṇi says that the *golādhyāya* contains six sections and the next chapter (that is, the *grahagaṇitādhyāya*) contains 10 sections, and it is furthermore clear that Cintāmaṇi considers that the *golādhyāya* constitutes the first part of the *Siddhāntasundara*.[70]

2. The *golādhyāya* opens with a number of invocatory and introductory verses that one would expect at the beginning of the work.[71] In contrast, the *grahagaṇitādhyāya* only has a single invocatory verse at its beginning.[72]

3. In the *grahagaṇitādhyāya* one finds references to information given in the *golādhyāya*. For example, one of the early verses of the *grahagaṇitādhyāya* states that the 47,400 divine years between the beginning of the *kalpa* and the commencement of planetary motion were mentioned earlier in the work.[73] While no previous mention of this is to be found in the *grahagaṇitādhyāya*, it is discussed in the *golādhyāya*.

It is thus established that the *golādhyāya* constitutes the first part of the *Siddhāntasundara*.

The question remains why the tradition became confused about the order of the two sections. The reason is probably that it is unusual to commence a traditional *siddhānta* with its *gola* section, and some readers and/or scribes must have felt that the *grahagaṇitādhyāya* was primary. It is, after all, in the *grahagaṇitādhyāya* that one finds the formulae that are needed for the actual practice of astronomy.

---

[70]The passage is quoted in an entry of a manuscript catalogue (see Velankar 1998, 1.4.94–95, entry 291).

[71]See *Siddhāntasundara* 1.1.1–3.

[72]See *Siddhāntasundara* 2.1.1.

[73]See 2.1.5.

Contents of the *Siddhāntasundara*

Based on the manuscripts used for the present translation of the *Siddhāntasundara*, the sections of the treatise are as given in Table I.2. The Sanskrit names of the sections are given, along with an English translation. It should be noted, however, that differences in the Sanskrit titles occur between the manuscripts. In the translation, mainly the English titles will be used, though occasional reference to the Sanskrit titles may occur.

It has already been noted (on page 25) that according to Jñānarāja's son Cintāmaṇi, the *golādhyāya* contains six sections, while the *grahagaṇitādhyāya* contains 10. In this regard, there are no problems with the *golādhyāya*, though the *grahagaṇitādhyāya*, as outlined in Table I.2, contains 11, not 10, sections. In fact, some manuscripts have 12 sections in the *grahagaṇitādhyāya*. In these manuscripts, the verses 14–23 from section 8 of the *grahagaṇitādhyāya* are repeated between sections 10 and 11, thus forming an independent section. However, in the present translation, the verses are kept in section 8 and not repeated later.

I.2.3 Jñānarāja's sources

Right at the beginning of the *Siddhāntasundara*, Jñānarāja explicitly gives his main source, as well as other treatises considered authoritative by him:[74]

> The topmost knowledge concerning the nature of the motion of the planets and the stars that was related to Nārada by the four-faced [Brahmā] was written down in its entirety by the sage bearing the name Śākalya. In my own verses, I am presenting precisely that [knowledge], accompanied by demonstrations (*vāsanā*s).
>
> Eight *tantra*s were written by [the gods and sages] Brahmā, Sūrya, Candra, Vasiṣṭha, Romaka, Pulastya, Bṛhaspati, and Garga. The difficult method of planetary computations is [given] in them. For crossing over the ocean [of this difficult science] by means of demonstrations [of the formulae] of their [that is,

[74] See *Siddhāntasundara* 1.1.3–4.

golādhyāya	1	*bhuvanakośādhikāra* Lexicon of the Worlds
	2	*bhuktihetu* Rationale of Planetary Motion
	3	*chedyakayukti* Method of Projections
	4	*maṇḍalavarṇana* Description of the Great Circles
	5	*yantramālā* Astronomical Instruments
	6	*ṛtuvarṇana* Description of the Seasons
gaṇitādhyāya	1	*madhyagatisādhana* Mean Motion
	2	*sphuṭagatisādhana* True Motion
	3	*tripraśna* Three Questions (on Diurnal Motion)
	4	*parvasambhūti* Occurrence of Eclipses
	5	*candragrahaṇa* Lunar Eclipses
	6	*sūryagrahaṇa* Solar Eclipses
	7	*grahodayāsta* Rising and Setting of Planets
	8	*tārāchāyābhadhruvādya* Shadows of Stars, Constellations, Polestars, and So On
	9	*śṛṅgonnati* Elevation of the Moon's Horns
	10	*grahayoga* Conjunctions of Planets
	11	*pāta* Occurrence of Pātas

Table I.2: Contents of the *Siddhāntasundara*

> the *tantras*'] mine of jewels, boats [in the form] of *siddhāntas* were made by Bhojarāja, Varāhamihira, Jiṣṇu's son [Brahmagupta], Caturveda [Pṛthūdakasvāmin], Āryabhaṭa, and the good Bhāskara [II].

The eight astronomical treatises (here called *tantras*) referred to by Jñānarāja are considered by the tradition to be divine revelations by the gods and sages mentioned rather than works authored by human beings; rather, texts composed by humans function as an aid to understanding the revelations of the divine beings.

The *Brahmasiddhānta* of the *Śākalyasaṃhitā*

The Śākalya that Jñānarāja refers to in the first verse is the supposed author of a *Śākalyasaṃhitā*. The lecture of Brahmā to Nārada, which Śākalya is said to have recorded, contains the contents of an astronomical treatise bearing the title *Brahmasiddhānta*. According to the *Brahmasiddhānta*, the text is the second part of the *Śākalyasaṃhitā*;[75] no other parts of a *Śākalyasaṃhitā* are known.

Jñānarāja explicitly says that what he has put into his own verses and augmented by adding *vasanā*s is the astronomical knowledge that was spoken by Brahmā to the sage Nārada and written down by the sage Śākalya. This shows that Jñānarāja's main source for composing the *Siddhāntasundara* is the *Brahmasiddhānta*.

The tradition of the eight *siddhānta*s

The *Brahmasiddhānta* mentions eight personages as the authorities from whom the science of astronomy originated:[76]

> This science [of astronomy] has come forth in eight ways: from me [Brahmā], from Candra, from Pulastya, from Sūrya, from Romaka, from Vasiṣṭha, from Garga, and from Bṛhaspati.

What is alluded to here is a tradition that the science of astronomy was originally given to mankind by eight deities and

[75] See Pingree 1970–94, A.4.259.
[76] *Brahmasiddhānta* 1.9–10.

ancient sages in eight treatises. Jñānarāja names the same eight deities and ancient sages as the originators of eight *tantras*: (1) Brahmā, (2) Sūrya, (3) Candra, (4) Vasiṣṭha, (5) Romaka, (6) Pulastya, (7) Bṛhaspati, and (8) Garga.

For the identification of the eight treatises, the following observations can be made: The first of them, of course, is the *Brahmasiddhānta*. The next four can be identified as follows: The lecture on astronomy by Candra (also known as Soma) is the treatise entitled the *Somasiddhānta*; the lecture by Vasiṣṭha is the treatise entitled the *Vasiṣṭhasiddhānta*; the lecture by Sūrya is the treatise entitled the *Sūryasiddhānta*; and the lecture by Romaka is the treatise entitled the *Romakasiddhānta*.[77]

The titles of the remaining three treatises would then, judging by the speakers, be *Pulastyasiddhānta*, *Bṛhaspatisiddhānta*, and *Gargasiddhānta*. However, these are rather doubtful. They may never have existed, or not have been preserved to the present.

The number eight is significant in Indian history. There are, for example, eight limbs in the system of *yoga*, eight limbs in the system of *āyurveda* (traditional Indian medicine), and eight *aṣṭaka*s (literally, eighths) of the *Ṛgveda* (an ancient collection of hymns sacred in Hinduism). It is therefore possible that the three doubtful treatises, the *Pulastyasiddhānta*, the *Gargasiddhānta*, and the *Bṛhaspatisiddhānta*, were added to the list to bring the number of treatises up to the significant number eight.

Other sources for the *Siddhāntasundara*

Other than the *Brahmasiddhānta* and the other related treatises, Jñānarāja mentions a number of other writers of astronomical texts:

[77] With the exception of the *Sūryasiddhānta*, for which an English translation and study are available (see Burgess 1858–60), none of these texts have been studied by scholars. The Sanskrit text of the *Brahmasiddhānta* has been published in Dvivedi 1912 and Dhavale 1996, the latter critically edited, and the Sanskrit text of the *Vasiṣṭhasiddhānta* has been published in Dvivedī 1907. The *Romakasiddhānta* is only available in manuscript form. It should be noted that there is sometimes more than one treatise with the same name. Varāhamihira, in his *Pañcasiddhāntikā*, describes five astronomical works, among which are a *Sūryasiddhānta*, a *Pauliśasiddhānta*, a *Romakasiddhānta*, and a *Vasiṣṭhasiddhānta* (see Pingree 1981, 11). These are not the same treatises as those referred to by Jñānarāja. For this reason, the *Sūryasiddhānta* relevant to us is often referred to as the modern *Sūryasiddhānta*.

1. Āryabhaṭa, a famous Indian astronomer who was born in 476 CE, is the author of the *Āryabhaṭīya* and the originator of the *āryapakṣa* and the *ārdharātrikapakṣa*;[78]

2. Brahmagupta, another famous Indian astronomer born in 598 CE and author of the *Brāhmasphuṭasiddhānta*;[79]

3. Bhojarāja, who flourished in the first half of the eleventh century CE and is the author of a now lost *Ādityapratāpasiddhānta* and a *karaṇa* entitled *Rājamṛgāṅka*;[80]

4. Varāhamihira, who flourished around 550 CE and wrote extensively on astronomy and astrology;[81]

5. Bhāskara II, who was born in 1114 CE and the author of many important works on astronomy and mathematics;[82]

6. Pṛthūdakasvāmin, who flourished in 864 CE and wrote an important commentary on Brahmagupta's *Brāhmasphuṭasiddhānta*;[83] and

7. Dāmodara.[84]

Although Āryabhaṭa, Bhojarāja, Brahmagupta, and Varāhamihira are mentioned as examples of authors of astronomical treatises, there are no indications that their works were used by Jñānarāja while composing the *Siddhāntasundara*; he presumably mentioned them owing to their status in Indian astronomy. Jñānarāja must have read the *Brāhmasphuṭasiddhānta* of Brahmagupta, though, for the commentary by Pṛthūdakasvāmin on this text is cited.[85] Jñānarāja clearly read the *Siddhāntaśiromaṇi* of Bhāskara II. Both Bhāskara II and the *Siddhāntaśiromaṇi* are mentioned by name, and the *Siddhāntaśiromaṇi*

[78]See Pingree 1970, Pingree 1970–94, A.1.50–53, A.2.15, A.3.16, A.4.27–28, A.5.16–17, and Plofker and Knudsen 2008a).

[79]See Pingree 1970–94, A.4.254–257, A.5.239–240. The editions of the *Brāhmasphuṭasiddhānta* used are Dvivedin 1902 and Ikeyama 2002.

[80]See Pingree 1970–94, A.4.336–339, A.5.266–267.

[81]See Pingree 1970–94, A.5.563–595, Pingree 1976a, and Plofker and Knudsen 2008b.

[82]See Pingree 1970–94, A.4.299–326, A.5.254–263.

[83]See Pingree 1970–94, A.4.221–222, A.5.224). Parts of the commentary on the *Brāhmasphuṭasiddhānta* have been edited (see Ikeyama 2002).

[84]See Dikshit 1969–81, 2.125–127 and Pingree 1978a, 614.

[85]See 1.1.23.

is cited.[86] In the *Siddhāntasundara*, Jñānarāja cites some *bīja* corrections from Dāmodara (see 2.1.83–84 and the commentary thereon).

Islamic influence in the *Siddhāntasundara*

As we have seen, the region in which Jñānarāja lived had been under Islamic control for a long time when the *Siddhāntasundara* was written.

Prior to the time of Jñānarāja, Islamic influence on Indian astronomy was either computational (that is, the only influence was on computation in formulae), as seen in the works of Muñjāla (fl. about 932 CE) and Śrīpati (fl. between 1039 and 1056 CE), or related to the introduction of the astrolabe into western India in the fourteenth century CE.[87]

In a survey of sixteenth-century Sanskrit astronomical treatises, including the *Siddhāntasundara*, from western and northern India, Pingree finds no reflections of Islamic astronomy, and he further notes that the next wave of Islamic influence on Indian astronomy occurred under the Moghuls, after the time of Jñānarāja.[88] The question remains, though, whether a careful investigation of the *Siddhāntasundara* would reveal Islamic influence.

On the whole, Pingree's assessment appears to be correct. Jñānarāja follows the Indian tradition when it comes to both mathematical formulae and theories about the cosmos. There are, however, two passages of interest, one dealing with what we today call Heron's method for finding square roots, the other with a cosmological idea involving a crystalline sphere.

Heron's method and iterative methods

Twice in the *Siddhāntasundara*, Jñānarāja gives a verse containing a method for computing the square root of any given positive number, once in the *grahagaṇitādhyāya*[89] and once in the *bīja*-

[86] See, for example, 1.1.32.

[87] See Pingree 1978b, 317–318. For Muñjāla, see Pingree 1970–94, A.4.435–436, A.5.312. For Śrīpati, see Pingree 1981, 25.

[88] See Pingree 1978b, 319.

[89] See 2.2.22.

gaṇitādhyāya.[90] The verse, as found in the *grahagaṇitādhyāya*, reads,

> An [approximate value of the desired] square root is increased by the result of the division of the given square [that is, the number whose square root we are seeking] by the approximate square root. [The result] is divided by 2. That is the [new] approximate square root. Then [the process is repeated] again and again. In this way, the correct square root [is found].

For the mathematical explanation of the verse, the reader is referred to the commentary on it. Here we will restrict ourselves to its historical significance.

The method described in the verse is an ancient one given by Heron of Alexandria in his *Metrica*;[91] we will refer to it in the following as Heron's method. Heron's method allows one to compute the square root of any given positive number by means of an initial approximation and iteration. In Jñānarāja's description of the method, iteration is prescribed until the square root is "exact." The verse has often been mentioned in modern studies,[92] and from modern surveys of methods for computing square roots in India, it is found that the *Siddhāntasundara* is the first Indian treatise to record the method.[93] Before Jñānarāja, the methods given in various mathematical works were not based on iteration, but were closed formulae.

Noting that Jñānarāja flourished at a time when Islamic culture penetrated into India, Chakrabarti suggests that Jñānarāja borrowed the method "from the Arabs."[94] Regarding Islamic involvement with the method, Smyly, without giving any references, says that the method "was known by the Arabs, but seems to have been subsequently forgotten,"[95] and Smith notes

---

[90] See Jain 2001, 13–14 for the details on the occurrence of the verse in one manuscript of the *bījagaṇitādhyāya* (Berlin 833).

[91] See Berggren 2002, 35.

[92] See, e.g., Datta and Singh 1935–38 (where an English translation is given, but no Sanskrit or other reference) and Bag 1979, 101 (Sanskrit given with a mistake). Note that Datta and Singh seem to have been first to cite the verse, and later studies refer to them.

[93] See Datta 1927, Datta and Singh 1993a, and Datta and Singh 1993b.

[94] See Chakrabarti 1934, 56.

[95] See Smyly 1944, 18.

that Rhabdas[96] followed an "Arabic method" when computing a square root using Heron's method.[97] However, there is no statement to the effect that the method is "Arabic" in the source referred to by Smith.[98] In fact, Youschkevitch, in his survey of Arabic mathematics, does not mention any methods for extracting square roots based on iteration,[99] and Berggren says that he is not aware of evidence that Islamic writers used Heron's method.[100]

In conclusion, it is highly unlikely that Jñānarāja learned Heron's method from Islamic sources. First of all, as shown above, this particular method was not used by Islamic mathematicians, and secondly, formulae for computing square roots as well as iterative methods have been known in India since early times.

Crystalline spheres

There is a passage in the *Siddhāntasundara* in which Jñānarāja potentially betrays a knowledge of foreign cosmology:[101]

> Some people say that for [each of] the moon, Mercury, Venus, the sun, Mars, Jupiter, Saturn, and the stars there are spheres rotating about the earth that are transparent like crystal. According to this opinion, on account of the fixed positions of the planetary orbits, their [that is, the planets' and stars'] rotation, [including] the perigees, apogees, nodes, and so on, is fixed by the polestar. It is not our opinion due to the weight of their forms.

[96]Nicolas Rhabdas was a Byzantine mathematician who flourished in the fourteenth century CE (see Berggren 2002, 35). Chakrabarti, curiously, writes "Kabdās" instead of "Rhabdas" in Chakrabarti 1934, 56.

[97]See Smith 1951–53, 254–255.

[98]Smith's source is Tannery 1886, 185.

[99]See Youschkevitch 1976, 76–80.

[100]See Berggren 2002, 35. Berggren further says that Rhabdas and other Byzantine mathematicians possibly learned Heron's method from Heron's work. It may be worth noting as well that Berggren reports that Al-Qalaṣādī, a fifteenth-century Muslim mathematician from Granada, recommends iterating the approximation $\sqrt{a^2+r} = a + \frac{r}{2\cdot a}$ once when $r > a$, yielding the new approximation $\sqrt{a^2+r} = a + \frac{r}{2\cdot a} - \frac{(r/(2\cdot a))^2}{2\cdot(a+r/(2\cdot a))}$ (see Berggren 2002, 34).

[101]*Siddhāntasundara* 1.2.9. The passage is the topic of Knudsen 2009.

In the verse, Jñānarāja tells us that some people hold the idea that for each of the planets and for the stars there are spheres that are transparent like crystal rotating around the earth.[102] Using weight as his argument, Jñānarāja rejects this idea of crystalline spheres.

Who these "some people" are is not revealed by Jñānarāja, and, unfortunately, Cintāmaṇi's commentary on the verse is not very helpful. Cintāmaṇi merely says, *spaṣṭārtham*, "The meaning is clear."[103]

Although neither Jñānarāja nor Cintāmaṇi identifies their source(s), it is clear that the idea expressed in the verse is derived from a foreign tradition. First of all, the idea of crystalline spheres is not found in the Indian tradition. Secondly, the idea of a crystalline sphere has been identified as being *yavana* (originally an Ionian or Greek, but later a person from the regions west of India) by Nṛsiṃha in his *Vāsanāvārttika* commentary on Bhāskara II's *Siddhāntaśiromaṇi*, and Pingree identifies *yavana* as Islamic.[104] Whatever the case may be, it is clear that the Indian tradition itself considered this idea to have come from outside. Regarding Nṛsiṃha's discussion of the idea, to be more precise, Nṛsiṃha argues that the *yavana* idea that there is a crystalline sphere supporting the sphere of the constellations and enabling it to rotate daily from east to west is incorrect. His argument against it is that the crystal could not bear such a weight.

It is worth noting that the word translated above as "opinion" is the Sanskrit word *pakṣa*. While it may mean "an opinion," it is also used to refer to the Indian schools of astronomy, such as the *brāhmapakṣa* or the *saurapakṣa*. It is not clear which of the two meanings the word refers to here, but there does not seem to be any Indian school of astronomy, or members of such, which subscribes to the idea of one or more crystalline spheres, so the intended meaning must therefore be "an opinion."

[102] When Jñānarāja says that the spheres are "transparent like crystal," this indicates that they are not made of crystal, but merely share the quality of transparency with crystal.

[103] Reading based on V_1 (f. 50v).

[104] See Pingree 1978b, 320–321, but note that the title of Nṛsiṃha's commentary is given incorrectly as Marīci in Pingree 1978b, 321, n. 34 (Marīci is the title of Munīśvara's commentary on the *Siddhāntaśiromaṇi*).

The problem is that there is no reference to a crystalline sphere or crystalline spheres in the Islamic literature.[105] The idea that one of the spheres (the ninth) is crystalline is a European idea, based on an interpretation of a passage in Ezekiel.[106] The word *yavana* could refer to Europeans, although a reference to Muslims is more likely. Still, this opens up the intriguing possibility that Jñānarāja was aware of a European cosmological idea.

I.2.4 Special features of the *Siddhāntasundara*

The *Siddhāntasundara* has the same general structure as older *siddhāntas*, especially the *Siddhāntaśiromaṇi* of Bhāskara II, but there are some unique features as well.

The *ṛtuvarṇana*

The *Siddhāntasundara* contains a *ṛtuvarṇana*, that is, a section containing a poetic description of the six seasons of the Indian tradition. It occurs as the sixth and last section of the *golādhyāya* (the Chapter on Cosmology) of the *Siddhāntasundara*.

As a poem in a technical treatise, all of the remaining sections of which deal with technical matters, the *ṛtuvarṇana* holds a special place. However, inclusion of a poem on the seasons in an astronomical treatise is not unique to Jñānarāja; Bhāskara II's *Siddhāntaśiromaṇi* contains a *ṛtuvarṇana* as well. Ultimately such sections are inspired by the *Ṛtusaṃhāra* of the famous poet Kālidāsa, sometimes called the Shakespeare of India.

Jñānarāja begins the *ṛtuvarṇana*, his poem about the seasons, with a description of spring in verses 2–13, and then he continues with the hot season in verse 14, the rainy season in verses 15–19, autumn in verses 20–29, early winter in verses 30–31, and cold winter in verses 32–33. With an introductory verse and a concluding verse, there are a total of 34 verses. In addition to the poem proper, there is a 35th verse marking the end of the section, as is the case for every section in the *Siddhāntasundara*.

In parts of the *ṛtuvarṇana* section, Jñānarāja employs poetic techniques of the tradition of Sanskrit poetry to create two registers of meaning in the text. That is, in places the text

[105] Personal communications with Jamil Ragep and others.

[106] See Grant 1987, 160–162 for the crystalline sphere in Europe.

expounds two narratives simultaneously.[107] One of these techniques is known as *śleṣa*, which we can loosely translate as *double entendre*, or "double meaning." Drawing on words with multiple meanings, two narratives are created in each verse where the technique is used. For example, the verses describing spring and the rainy season simultaneously describe the activities of the god Kṛṣṇa in those seasons. Similarly, the verses describing fall also describe the activities of the god Rāma in that season.

The translation and commentary of the *ṛtuvarṇana* in the present book are a slightly revised version of those by Minkowski and Knudsen.[108] A general overview of seasonal poetry in Sanskrit astronomical treatises has been done by Minkowski.[109]

Other instances of poetical techniques

Jñānarāja's use of *śleṣa* is not confined to the *ṛtuvarṇana* section, but is also found in some verses in section 3 of the *grahagaṇitādhyāya* (Chapter on Mathematical Astronomy). There one layer of meaning is narrative, the other technical. More specifically, the narrative layer gives a story and poses a question that cannot be answered from the story alone; the technical layer provides the information necessary to answer the question posed.

The nature of these verses is such that it is not possible to capture their meaning in one English translation. Two translations, one narrative and one technical, are needed.

Problems are not otherwise found in *siddhānta*s, so the presence of them is unique. In addition, the complex poetry that is used to phrase them makes them stand out even more.

This is best illustrated by an example, for which we will use *Siddhāntasundara* 2.3.17:

> [Narrative translation:] Tell [me] the length of the journey of the swift man, who, upon learning that his friend had gained kingship and was sitting on the lion's seat [that is, the throne], deprived of luster went north [to a place where] he had the luster of a former king.

[107] See Bronner 2010 for a recent study of double narratives in Sanskrit literature.

[108] See Minkowski and Knudsen 2011.

[109] See Minkowski 2011.

> [Technical translation:] Tell [me] the measure of the journey of the swift man, who, upon learning that the sun had attained lordship sitting in the lion's seat [that is, was in Leo], cast no shadow and who went north [until] he had a shadow of [length] 16 [falling] toward the east.

As can be seen, the narrative itself does not provide the information needed to solve the question posed; to solve it, the words of the verse have to be interpreted differently, yielding the technical translation given, which provides the necessary information for the solution of the problem. For the technical aspects of the verse and the solution to the problem, the reader is referred to the translation and commentary on page 226.

As a further example, consider the first quarter of another such verse (*Siddhāntasundara* 2.3.19):

> [Narrative translation:] When the good goose perished in the mouth of a crocodile...
>
> [Technical translation:] When the sun, being in the beginning of Capricorn, was setting...

I.2.5 Importance of the *Siddhāntasundara*

As already noted, the *Siddhāntasundara* was the first *siddhānta* to be written after Bhāskara II composed the *Siddhāntaśiromaṇi* in 1150 CE. In the intervening period, the Indian astronomers moved in a didactic direction. A number of minor *siddhānta*s, such as the *Brahmasiddhānta*, and the *Somasiddhānta*, all of which were discussed above, were composed in this period, but no important *siddhānta* was written.[110] In other words, Jñānarāja's *Siddhāntasundara* was the first major *siddhānta* to be written after the influential *Siddhāntaśiromaṇi*.

Jñānarāja and the astronomical tradition

Writing the first major work after some 350 years naturally entails a critical look at and a reassessment of the tradition. This is so in the case of Jñānarāja, and a direct statement as to what shape such a reassessment took in the *Siddhāntasundara* can be

[110] See Pingree 1981, 26.

found right at the beginning of the text. Jñānarāja says there that what he has written in the *Siddhāntasundara* is the astronomical teachings of Brahmā to Nārada, as recorded by the sage Śākalya, with *vāsanā*s added. Two significant points regarding Jñānarāja's motivations for writing the *Siddhāntasundara* are raised here:

1. Jñānarāja takes as his main source the somewhat obscure *Brahmasiddhānta*; and

2. his contribution to astronomy is his addition of demonstrations to the science as it is given in the *Brahmasiddhānta*.

The *Brahmasiddhānta*, as well as related texts such as the *Somasiddhānta*, are considered authoritative by Jñānarāja as a result of their being authored by divine beings and ancient sages. By accepting the *Brahmasiddhānta* as his primary source, Jñānarāja emphasizes, in addition to emphasizing the *Brahmasiddhānta* itself, the fact that it is attributed to a deity. This emphasis is further strengthened when he makes a clear division between the eight divine originators of astronomy and the human beings who subsequently wrote treatises on the topic. The divine works are primary, human works secondary. However, as per the second point above, while Jñānarāja makes no claims to being anything but an ordinary human being, he argues that he is making a contribution, namely, adding *vāsanā*s.

What all this implies is that Jñānarāja is seeking to bring the *Brahmasiddhānta* and similar treatises into the mainstream of Indian astronomy. As was discussed before, with the exception of the *Sūryasiddhānta*, the eight treatises mentioned by Jñānarāja are somewhat obscure. By composing a treatise emphasizing these texts, in particular the *Brahmasiddhānta*, and providing *vāsanā*s and rationales that they are lacking, the texts are drawn into the mainstream of Indian astronomy, to be discussed on the same level as, say, Bhāskara II's *Siddhāntaśiromaṇi*, which has *vāsanā*s.

Why is Jñānarāja following the *Brahmasiddhānta* in particular? No answer is given by Jñānarāja in the *Siddhāntasundara*, but we may venture an explanation here. As we will see below on page 40 regarding the *virodhaparihāra* issue, Jñānarāja is concerned with traditional religious teachings. He is willing to reinterpret passages from sacred texts, but not willing to question their authority. Dikshit notes that the *Brahmasiddhānta* treats the subject of religion, a subject not met with in astronomical texts.[111] Unfortunately, however, he does not specify this further. No in-depth study of the *Brahmasiddhānta* is available, but it seems likely that this particular text matched Jñānarāja's agenda, or perhaps that Jñānarāja was inspired by its approach.

Relationship with Bhāskara II

While Bhāskara II's *Siddhāntaśiromaṇi* is in many cases a model for the *Siddhāntasundara*, Jñānarāja is critical of Bhāskara II in many places. The reason is probably that Jñānarāja, writing a *siddhānta* following the *Siddhāntaśiromaṇi*, felt the need to justify his work. On the one hand, he was perpetuating a tradition, but on the other, he had to critique the tradition to justify his own contribution.

As an example of Jñānarāja vying with Bhāskara II, Jñānarāja's *ṛtuvarṇana* contains 34 verses, more than double the 15 verses in Bhāskara II's *ṛtuvarṇana*, and it employs more elaborate poetic techniques.

Primacy of the *golādhyāya*

We established previously that the *Siddhāntasundara* begins with the *golādhyāya* (Chapter on Cosmology). In earlier *siddhānta*s, such as the *Śiṣyadhīvṛddhidatantra*, the *Vaṭeśvarasiddhānta*, and the *Siddhāntaśiromaṇi*, the cosmology chapter is placed at the end of the work.[112] In addition, the authors of these works found it necessary to preface their cosmology chapters with arguments for why this section is to be studied.[113]

In contrast, Jñānarāja does not find it necessary to justify his *golādhyāya*; he simply begins with it. Importance was put on the cosmology material by all the great Indian astronomers, but to Jñānarāja it was central: a study of astronomy commenced with a study of the *golādhyāya*. This, of course, aided his endeavor to add *vāsanā*s to the material in the *gaṇitādhyāya*, and also gave

---

[111]See Dikshit 1969–81, 2.49.

[112]At least this is how the printed editions organize the texts (see Chatterjee 1981, Shukla 1986, and Śāstrī 1999).

[113]See, e.g., *Śiṣyadhīvṛddhidatantra* 24.2–6, *Vaṭeśvarasiddhānta* (*gola*) 1.1–5, and *Siddhāntaśiromaṇi* (*golādhyāya*) 1.2–5.

his readers the theoretical understanding necessary to engage with questions such as that of *virodhaparihāra*.

The issue of *virodhaparihāra*

By the Sanskrit term *virodhaparihāra* (literally, "removal of contradiction") is meant the endeavor to create a synthesis between the cosmology of the *purāṇas* on the one hand and the cosmology of the astronomical tradition on the other. The astronomers' approach to the inconsistencies between the two cosmologies was to incorporate compatible elements into their own system, while rejecting contradictory elements. Thus, Meru, the world mountain, found its place on the North Pole of the spherical earth and the annular oceans and continents were placed on the Southern Hemisphere, while notions such as the earth being flat and the sun being closer than the moon were rejected.

The approach changed with Jñānarāja. Rather than accepting certain elements and rejecting others, Jñānarāja considered the *purāṇas* authoritative and sought to create a synthesis in which both cosmologies could be true. This entailed reinterpreting ideas from both traditions and rejecting some ideas of the astronomical tradition, but never rejecting any elements from the *purāṇas*.

This approach of Jñānarāja is expressed in section 1 of the *golādhyāya* (Chapter on Cosmology), the very first section of the *Siddhāntasundara*.[114]

I.3 Textual basis of this translation

There is an edition of the Sanskrit text of the *Siddhāntasundara* available, namely, that done by Śukla.[115] However, even though multiple manuscripts were consulted for the edition, it is not a critical edition, nor is it even carefully edited. A critical edition of section 1 of the *golādhyāya* (Chapter on Cosmology) and sections 1–6 of the *grahagaṇitādhyāya* (Chapter on Mathematical Astronomy) can be found in Knudsen 2008. A complete critical edition of the *Siddhāntasundara* is currently under preparation by the present author.

[114] See also Minkowski 2004a for a discussion.

[115] See Śukla 2008.

The translation of the *Siddhāntasundara* presented in this book is based on 20 manuscripts, photocopies of which were made available by the late David Pingree:[116]

1. Ānandāśrama 4350. Catalogue description: not available.
2. Bhandarkar Oriental Research Institute 283 of Viśrāma (i).[117]
3. Bhandarkar Oriental Research Institute 219 of A 1882–83.[118]
4. Bhandarkar Oriental Research Institute 880 of 1884–87.[119]
5. Bhandarkar Oriental Research Institute 881 of 1884–87.[120]
6. Bhandarkar Oriental Research Institute 860 of 1887–91.[121]
7. India Office Library 2114b.[122]
8. British Museum Add. 14,365p.[123]
9. Asiatic Society of Bombay 289.[124]
10. Asiatic Society of Bombay 290.[125]
11. Asiatic Society of Bombay 291.[126]
12. Oxford d. 805(5).[127]
13. Rajasthan Oriental Research Institute (Kota) 981.[128]

[116]For more details on each manuscript, see Knudsen 2008, 50–59 or the upcoming complete critical edition of the *Siddhāntasundara* by the same author.
[117]Catalogue description: Navathe 1991, 350–351, no. 1240.
[118]Catalogue description: Navathe 1991, 349–350, no. 1239.
[119]Catalogue description: Navathe 1991, 352, no. 1242.
[120]Catalogue description: Navathe 1991, 351, no. 1241.
[121]Catalogue description: Navathe 1991, 352–353, no. 1243.
[122]Catalogue description: Eggeling 1886–1904, 1029, entry 2902.
[123]Catalogue description: Bendall 1902, 187, no. 452.
[124]Catalogue description: Velankar 1998, 1.4.94, no. 289.
[125]Catalogue description: Velankar 1998, 1.4.94, no. 290.
[126]Catalogue description: Velankar 1998, 1.4.94–95, no. 290.
[127]Catalogue description: Pingree 1984, 8, no. 21.
[128]Catalogue description: Trivedi and Sharma 1992, 484–485, no. 4306.

14. Rajasthan Oriental Research Institute 4733.[129]

15. Rajasthan Oriental Research Institute (Jodhpur) 37247.[130]

16. Benares 35318.[131]

17. Benares 35566.[132]

18. Benares 35627.[133]

19. Benares 36902.[134]

20. Benares 36907.[135]

The translation work has been challenging, and while depth has been reached in many places, other passages still appear lacking. It is the hope that the scholarly community will be forgiving of whatever errors there are, and that they may be corrected by the next generations.

[129] Catalogue description: Jinavijaya 1965, 310–311, no. 5535.

[130] Catalogue description: Sharma and Singh 1984, 344–345, no. 3079.

[131] Catalogue description: Staff of the Manuscripts Section 1963, 96–97, no. 35318.

[132] Catalogue description: Staff of the Manuscripts Section 1963, 120–121, no. 35566.

[133] Catalogue description: Staff of the Manuscripts Section 1963, 124–125, no. 35627.

[134] Catalogue description: Staff of the Manuscripts Section 1963, 242–243, no. 36902.

[135] Catalogue description: Staff of the Manuscripts Section 1963, 242–243, no. 36907.

1. Chapter on Cosmology

Section 1: *Lexicon of the Worlds*

The first section of the *Siddhāntasundara*, which is also the first part of the Chapter on Cosmology in the work, is entitled *Lexicon of the Worlds*. In many ways, this section is the most important one in the *Siddhāntasundara*, as it lays out many of the ideas held by its author, Jñānarāja, including *virodhaparihāra*.

The section gives three different accounts of creation, various philosophical arguments, terrestrial geography, and cosmology.

(1) **I praise [the god Gaṇeśa, who is of] auspicious form, who is primeval and imperishable, who is dear to his devotees [or, to whom his devotees are dear], on whose forehead the moon is [resting], on whose temples a line of black bees are congregated in search of nectar, on whose throat a serpent is shining fervently, at whose two feet the hearts of the gods [dwell], and whom even [the creator-god] Brahmā, desiring to create the three worlds, served for the sake of unhindered success.**

Jñānarāja opens the *Siddhāntasundara* with an invocation praising the elephant-headed god Gaṇeśa. The verse could also be taken to refer to the god Śiva, although the reference to bees on the temples of the deity seeking nectar seems to indicate Gaṇeśa, as the sweat from an elephant's forehead is considered to attract bees in the Indian tradition.

Gaṇeśa is traditionally considered to be the god that removes obstacles, and scribes copying Sanskrit texts often open with invoking Gaṇeśa to ensure success in their undertaking. By invoking Gaṇeśa at the beginning of his treatise, he seeks success in completing it. It should also be noted here that Jñānarāja described his father Nāganātha as a devotee of Gaṇeśa,[136] which Jñānarāja might have been as well.

(2) **I, Jñānarāja, am composing the *Siddhāntasundara*, which gives bliss to the learned, and which is ingenious and correct, after bowing down to my teachers as well as to [the goddess] Bhuvaneśvarī, [whose] worshipper [though filled with] inner night may become one who has destroyed [his] darkness and [then] possessed of increasing arts through the ray-like syllables of her name, which sound as they occupy his body [in a *yantra*, that is, an object or a symbol used for worship, or as they are deposited on the worshipper's body ritually] in sequence.**

In this verse, there is a distinct flavor of Tantrism, a term covering a number of esoteric traditions in India whose teachings are given in texts known as *tantra*s (not to be confused with the *tantra*s of the astronomical tradition). Here Jñānarāja invokes Bhuvaneśvarī, a Hindu goddess worshipped in certain forms of Tantrism.

(3) **The foremost knowledge concerning the nature of the motion of the planets and the stars that was related to Nārada by the four-faced [god Brahmā] was composed in its entirety by the sage bearing the name Śākalya. In my own verses, I am presenting precisely that [knowledge], accompanied by demonstrations.**

Here the source of the *Siddhāntasundara* is given as the *Brahmasiddhānta*, a treatise claiming to have been narrated by the deity Brahmā to the sage Nārada and recorded by the sage Śākalya.Śākalya. This has been discussed in the Introduction (see page 28).

(4) **Eight *tantra*s were written by [the three gods] Brahmā, Sūrya, [and] Candra, [and by the five sages] Vasiṣṭha, Romaka, Pulastya, Bṛhaspati, and Garga. The difficult method of planetary computations is [given] in**

[136] See Introduction, page 6.

them. For crossing over the ocean [of this difficult science] by means of demonstrations [of the formulae] of their [i.e., the *tantra*s's] mine of jewels, boats [in the form] of *siddhānta*s were made by [the astronomers] Bhojarāja, Varāhamihira, Jiṣṇu's son [Brahmagupta], Caturveda [Pṛthūdakasvāmin], Āryabhaṭa, and the good Bhāskara [II].

Jñānarāja here supports the tradition that the science of astronomy was received through eight revelations written down in eight treatises. These eight treatises are normally known as *siddhānta*s, but Jñānarāja here calls them *tantra*s, reserving the term *siddhānta* for the subsequent treatises written by human beings. The human astronomers mentioned are all well known in the Indian tradition.

(5) **The triad of *veda*s came forth for the sake of rites such as sacrifices and so on. For computing that which relates to direction and time [in the performance of these rites], this *śāstra* [that is, *jyotiḥśāstra*, more specifically astronomy], which is to be studied by the twice-born, was taught by the ancient sages.**

In the rites prescribed in the *veda*s and ancillary literature, correct timing and orientation with respect to the cardinal directions are very important. The field of astronomy provides the knowledge by which the correct time and direction can be determined, and as such plays an important role in the performance of the rituals of the *veda*s, sacred texts in Hinduism in which ritual plays a major role.

The term *jyotiḥśāstra* will be defined below in verse 8.

(6) **It is said by the ancient sages that he who begins his religious observances during any forbidden time, such as during the intermission of study, is not a twice-born. [And the ancient sages have also said that] a woman who has the same *ghaṭikā* is a sister [that is, she is not marriageable, and to marry her is to commit incest]. And an action [performed] in confusion about direction [that is, which direction is east] is fruitless, and so are obser-**

vances on the [special] *tithi*s and so on [that is, they are fruitless when performed in the wrong direction].

That a woman has the same *ghaṭikā* as a given man means that she has the same ascendant (the point on the horizon where the ecliptic is rising) in her birth chart as the man. In this case, the woman is to be considered a sister, and the marriage between the two is forbidden. However, whether this is the case can only be known through the science of astronomy, the importance of which is thus emphasized.

(7) **The lord of the gods in the form of the *veda* conquers to protect the world. That which is his mouth is *vyākaraṇa*; *nirukta* is said to be his ear; likewise, his nose is *śikṣā*; *jyotiṣa* is his eye; his pair of hands is *kalpa*; and his pair of lotus-like feet is *chandas*. In this way the six-limbed personified *veda* is to be understood with reference to meaning and recitation.**

The *vedāṅgas* (literally, limbs of the *veda*) are six auxiliary disciplines through which the *veda* can be properly understood. Together these disciplines are said to form the body of the *veda*, as it is through them that a person understands the *veda* and becomes able to practice its doctrines.

The first *vedāṅga* mentioned by Jñānarāja is *vyākaraṇa*, the science of grammar, which is considered the face of the embodied *veda*; the second, *nirukta*, the science of the etymology of the words of the *veda*, is considered the ear; the third, *śikṣā*, the science of correct pronunciation of the words of the *veda*, is considered the nose; the fourth, *jyotiṣa*, is considered the eye; the fifth, *kalpa*, the science of ritual, is considered the hands; and the sixth, *chandas*, the science of poetic meters in the *veda*, is considered the feet. Note that there is more than one meaning of the Sanskrit word *kalpa*. Here it refers to the science of ritual, but it also (as in verse 8 and verse 12) denotes a world age of duration 4,320,000,000 years (see Introduction, page 16). This span of time is the life span of the world. At the beginning of the *kalpa*, the world comes into being, and at its end it is destroyed. The *kalpa* corresponds to a day in the life of the creator-god Brahmā. The world ages will be explained in greater detail later in the treatise by Jñānarāja (see *Siddhāntasundara* 2.1.2–3 and the commentary thereon).

(8) ***Jyotiḥśāstra* has three divisions: astronomy, astrology, and omens. The computation of the planets [i.e., of planetary motion] is the foremost part of the [astronomical treatises called] *tantras*. A *siddhānta* is [a treatise] where the nature of the nectar relating to the planets as well as their motions and measures in a *kalpa* [is given], and computation is given as well with demonstrations.**

The division of *jyotiḥśāstra* given here is a traditional one,[137] dividing *jyotiḥśāstra* into astronomy, astrology, and omens.

Jñānarāja explains in more detail his distinction between *tantras* and *siddhāntas*: a *tantra* contains the computational matters, whereas a *siddhānta* contains demonstrations (*vāsanās*) as well.

When Jñānarāja speaks of the "nectar relating to the planets [and so on]," he is using what is a common metaphor in India. The meaning is either that the *siddhāntas* present the best of the science of astronomy, or that this science is nectar, that is, a high and important science.

(9–10) **First, the principle of intellect arose from a combination of original nature and self for the creation of the three [worlds]. The sense of "I" arose from this source. From that the basic element of sound arose. From that the sky arose. From that the basic element called contact arose. [From that] the wind arose. From the wind the basic element of form arose. From that light arose. From that the basic element of taste arose. [From that] water arose. From water the basic element of smell arose. From that earth arose. From the joining together of these this [world] arose.**

Sāṅkhya is one of the schools of classical Indian philosophy. It is a dualist philosophy that regards the world as being composed of two eternal elements: *puruṣa* (self) and *prakṛti* (nature).[138]

[137] See Pingree 1981, 1.
[138] See Scharf 2008.

The unfolding of creation according to this system of philosophy is here briefly outlined by Jñānarāja. It is the first of three creation accounts that he will present.

(11–12) **The universe is resting in the belly of the All-Creator. From his two feet the earth [came to be]; from his navel the [intermediate] region known as the atmosphere [came to be]; from his head the heaven [came to be]; from his mouth [the deities] known as Indra and Agni [came to be]; and the deity of wind [Vāyu] arose from his breath. Thus it is said in the *veda*. The moon is dwelling in his heart; the sun dwells in his eye; from his ear all the cardinal directions arose. This is the path of creation in the present *kalpa*.**

This account of creation is taken from a famous creation hymn, the *puruṣasūkta*, of one of the most sacred texts of Hinduism, the *Ṛgveda*.[139]

(13–19) **Awakened by the middle vital air, the seven vital airs, which merged into the Primeval Person [i.e., the god Viṣṇu] during the previous *pralaya* [destruction], created seven Persons. By them, having become one, the Expansive Person [i.e., Brahmā] was created. He appeared from a lotus. As all this [i.e., the universe] is fashioned by him, it is called his creation.**

At first, Brahmā created the waters from his own speech. [Then] he entered [the waters] along with the triple [sacred science] by means of a portion of himself. An egg arose. From [the embryo of] that [egg] Agni [the god of fire] arose here. After [Brahmā] pressed the pair of [half-]shells [of the egg] together and put them in the water, the earth was created. Then he joined with that [earth] possessing a portion of Agni. An egg arose, and Vāyu [the god of wind] arose from within that [egg].

At that very place, the atmosphere came to be from the shell of the egg. Joining with that [sky], having joined his own portion with Vāyu [the god of wind, or merely wind], he made an egg. This sun arose from

[139] *Ṛgveda* 10.90.13–14.

that [egg]. In that place, the heaven was [born] from the shell of that egg and the sunbeams [arose] from the juice sticking to the half-shell. Joining with that [sky] by means of his own portion [and] along with a portion of the sun, an egg appeared.

The moon arose from within that egg. The multitude of stars arose from the water flowing in this place. The cardinal directions [arose] from the half-shell of that [egg], and the intermediate directions [arose] in like manner from the matter sticking to that [shell].

Having created the worlds from the combination of speech and mind, the All-Creator created the eight [gods known as the] Vasus by means of eight months. Likewise, the [gods known as the] Rudras, as well as the 12 [solar gods known as the] Ādityas, and the [gods known as the] Viśve Devas.

The Creator placed Agni [the god of fire] and [the gods known as] the Vasus on the earth, [gods known as] the Maruts and the group of [gods known as] Rudras in the atmosphere, the Ādityas associated with the sun in the sky, and the moon accompanied by the Viśve Devas in the cardinal directions.

Thus it is taught in the text of the beginning of the sixth *kāṇḍa* [that is, section] of the *Śatapathabrāhmaṇa*. Here [in that work] a particular path of creation is given. There is oneness of both creation [accounts when examined] with laudable reflections. If [however] there is a difference in some place, it is to be understood through *kalpabheda*.

The most voluminous of the three creation accounts presented is this one from the *Śatapathabrāhmaṇa*.[140]

The term *pralaya* refers to the destruction of the universe. This is followed by a new creation.

The association of the eight Vasus with months is unclear; the *Śatapathabrāhmaṇa* has *drapsa*, "drop," instead of *māsa*, "month."[141]

The term *kalpabheda* (literally, different *kalpa*) is often used

[140] *Śatapathabrāhmaṇa* (Mādhyandina recension) 6.1.1.1–6.1.2.10.
[141] *Śatapathabrāhmaṇa* 6.1.2.6.

by the interpreters of the *purāṇas*. If inconsistencies occur in two accounts of the same story, they are explained by saying that the two accounts took place in two different world ages (*kalpas*). What this means is that certain events are considered to take place in every creation, that is, in every *kalpa*, but the details might differ. As such, inconsistencies in myths or religious narratives given in different sacred texts can be explained without invalidating any of them. Śrīdharasvāmin, the renowned commentator on the *Bhāgavatapurāṇa*, invokes *kalpabheda* twice.[142] Here, of course, the word *kalpa* refers to the world age of 4,320,000,000 years (see commentary on verse 7).

(20) **The circle of the earth, the oceans, the mountains, the planets, and so on were created, in order, in 474 times 100 divine years [i.e., 47,400 divine years] commencing at the beginning of the *kalpa*. Then all the planets were placed on the circle of stars.**

That a period of 47,400 divine years, during which the earth, planets, and so on are created, occurs at the beginning of the *kalpa* is a basic idea of the *saurapakṣa*.[143] Since a divine year is equal to 360 of our years, the period is equivalent to 17,064,000 years. This period is called *śṛṣṭikāla*, "creation-time," in Sanskrit. It should be noted that since the planets are being created during this period, there is no planetary motion during the *śṛṣṭikāla* (see also Siddhantasundara 2.1.7, which condemns the idea that planetary motion begins at the beginning of the *kalpa*).

(21) **The earth-sphere is indeed globular. The oceans are located as girdles on that [earth-sphere] on which, in the middle, on the top, and at the bottom, there are gods, men, demons, mountains, and trees.**

Just as the protuberance of a *kadamba* flower is supporting its filaments, so this preeminent [earth-sphere], which is holding [all these gods, men, and so on], rests immovable in space, its great weight supported by the *avatārapuruṣas*.

[142] Commentary on *Bhāgavatapurāṇa* 5.16.27–28 and 12.11.39 (the edition used is Shastri 1988). See also Minkowski 2004a, 355.

[143] See *Sūryasiddhānta* 1.24.

In the Indian tradition, there are a number of ring-shaped oceans situated on the earth as girdles (see verse 41 below). The *avatārapuruṣas* are the incarnations, or physical manifestations, of the god Viṣṇu that, according to the Indian tradition, hold up the earth.

The *kadamba* flower is the flower of the tree *anthocephalus cadamba*. It is shaped as a ball with tiny white petals that point in all directions. The analogy with the sphere of the earth, which has people, and so on, all over it, is clear.

(22–23) **Since a planet located [directly] above the city of Laṅkā is [due] south on the horizon of [the mountain] Meru, [due] west on the horizon for a person in Yamakoṭi, [due] east on the horizon in Romakapura, and [due] north on the horizon from Vaḍavānala, therefore the earth indeed has the form of a sphere.**

Since this good argument for the sphericity of the earth which Pṛthūdakasvāmin gives is not acquired by means of the *pramāṇas*, therefore low-minded people do not value it.

Laṅkā, Romakapura, Siddhapura, and Yamakoṭi are four imagined cities on the terrestrial equator, each being 90° from its two neighbors. Laṅkā is famous in Indian mythology as the capital of the demon-king Rāvaṇa, who was killed by thc god Rāma. Meru, the legendary mountain of Indian mythology, is on the North Pole, and Vaḍavānala is an underwater fire at the South Pole. Vaḍavānala is also known as Vaḍavāgni, the name used in verses 40 and 44–49.

An argument for the earth's sphericity, given by Bhāskara II and Pṛthūdakasvāmin,[144] is the following. When a planet is directly above Laṅkā, it is on the horizon for someone at Meru (note that the southern direction is not uniquely determined at Meru, and the northern direction is similarly not uniquely determined at Vaḍavānala), due east on the horizon for someone in Romakapura, due west on the horizon for someone in Yamakoṭi, and on the horizon for someone at Vaḍavānala. In other

[144] Pṛthūdakasvāmin presents this argument in his commentary on *Brāhmasphuṭasiddhānta* 21.1.

words, if we imagine a person standing on the North Pole, a planet directly above Laṅkā (or any other location on the celestial equator) will be on the person's horizon. Of course, this argument is only theoretical unless one is able to place people at specific places on the earth at a specific time, when a planet is at a specific location. In other words, this is not an argument that could be demonstrated in medieval India.

In Indian philosophy, the *pramāṇa*s are ways of acquiring knowledge. The number of *pramāṇa*s varies in different texts, but the list generally includes direct observation, logical inference, and supernatural authority. That the argument given in the verse is weak owing to its not being based on the *pramāṇa*s is explicitly noted by Jñānarāja,[145] though he implies that only a low-minded person would need an actual demonstration.

(24) **For a traveler who is facing the polestar [that is, traveling north or south] or facing horizontally [that is, traveling east or west], for each *yojana* [traveled] the earth is said to create, in order, an elevation of the nonmoving [polestar] situated far from the earth, or to produce the comparable experience of sameness [of elevation]. For this very reason, it is like a ball.**

The idea of the verse is that if one travels east or west, no change in altitude (elevation) above the horizon is perceptible for the polestar is seen, whereas a change in altitude is seen when traveling north or south.

For the *yojana*, a unit of length, see Introduction, page 17.

(25) **For a traveler who is** 14 ***yojana*s north of his own region the circle of the *nakṣatra*s [i.e., the ecliptic], which is directed toward the polestar, is depressed from the zenith toward the south, and the polestar has an altitude of** 1 **degree above the horizon.**

According to the verse, when traveling 14 *yojana*s north, the polestar is elevated 1°. Let c be the circumference of the earth

[145] For the defects of the argument and Cintāmaṇi's treatment of it, see Minkowski 2002, 505.

in *yojanas*. Since c corresponds to $360°$, we get the proportion

$$\frac{1}{14} = \frac{360}{c}, \tag{1}$$

which yields

$$c = 14 \cdot 360 = 5{,}040. \tag{2}$$

In other words, the numbers given in the verse imply that the circumference of the earth is 5,040 *yojanas*. For this result, see also *Siddhāntasundara* 2.3.18. Later in this section (verse 74), however, the circumference will be given as 5,059 *yojanas*. In the next verse, a thought experiment yields the circumference of the earth as 5,000 *yojanas*.

(26) **When the sun was on the eastern horizon [that is, when it was rising], a swift-moving man commenced an eastward journey of 10 *yojanas* holding a sand clock in his hand. Learning that the time [of sunrise at his destination] was 7;12 *palas* less than sunrise at his origin [on the day before], the earth was understood by him to indeed be in the form of a sphere measuring 5,000 [*yojanas* in circumference].**

Here Jñānarāja presents a thought experiment. At sunrise, a man carrying a sand clock in his hand travels 10 *yojanas* eastward. When the next sunrise occurs, he notes that it does so 7;12 *palas* (a unit of time equal to $\frac{1}{60}$ *ghaṭikā*) earlier than on the previous day. Since this difference in time corresponds to the 10 *yojanas* that he traveled, the man is able to compute the circumference of the earth.

The text literally says *seṣvaṃśādripalādhikaṃ tu samayaṃ*, "time greater by 7;12 *palas*," but the meaning must be that sunrise occurred 7;12 *palas* earlier at his destination than at his origin on the previous day, as reflected in the translation.

The value of the circumference is derived from a simple proportion. Let c be the circumference of the earth. There are 3,600 *palas* in a nychthemeron (that is, a day and a night, or 24 hours), and so

$$\frac{c}{10} = \frac{3{,}600}{7;12}, \tag{3}$$

which gives us that

$$c = \frac{10 \cdot 3{,}600}{7;12} = 5{,}000. \qquad (4)$$

Note that this holds true only if the man was traveling along the terrestrial equator.

(27) **[Even] after seeing the word *bhūgola* [that is, earth-sphere] being used in the *purāṇas* as well as the phrase "Meru is north of all [places]," people who are disposed to obstinacy say that the entire earth is [flat] like the surface of a mirror. But none of them know the meaning of the *purāṇas*, nor that the spherical nature of the earth is established by excellent demonstrations.**

The word *bhūgola*, which literally means "earth-sphere," is used five times in one of the most well known of the *purāṇas*, the *Bhāgavatapurāṇa*.[146] That it occurs in the *purāṇas* is the basis of Jñānarāja's argument that the author of these texts knew that the earth is spherical.

(28) **The likeness to the surface of a mirror mentioned in the *purāṇas* [applies only] to a one-hundredth part of the earth, not the [entire] sphere of the earth. A one-hundredth part of the circumference [of the earth] is seen [as being straight] like a stick. Therefore, the sphere of the earth appears as if it is flat to human beings.**

The statement of the *purāṇas* that the earth is flat is not to be taken literally. It merely reflects that a small section of it appears flat to a human being.

(29–30) **It is said in the treatise of Bhāskara [II] that the logical flaw of infinite regression arises when an embodied supporter of the earth is assumed, and that therefore firmness should be taken as an inherent quality of the earth, just as heat is a quality of fire and fluidity of water. That [argument] is not correct.**

[146] *Bhāgavatapurāṇa* 3.23.43, 5.16.4, 5.20.38, 5.25.12, and 10.8.37.

Let Śeṣa and the others, who are mentioned in the *purāṇa*s, be holding steady the motion of the earth. Since they are mentioned in the *veda*s regarding the motion of the planets and the cage of the stars, what is the fault with them [as being the support of the earth]? Moreover, if there is a quality characterized by firmness in the earth, [then] why [is that] not [so] in its parts, just as there is fluidity in drops of water, heat in sparks of fire, and so on?

According to Bhāskara II, if we assume a support of the earth, we end up with an infinite regress. For what is the support of the support, and the support of the support's support? It is better, says Bhāskara II, to assume that the earth has an inherent quality of firmness, just as fire has the quality of heat and water that of fluidity. In other words, the earth is its own support.[147]

Jñānarāja disagrees. If the earth has the quality of firmness, why is this not so for its parts? If we place a piece of earth in the air it will fall. However, even small drops of water and sparks of fire have the respective qualities of fluidity and heat.

Śeṣa is the incarnation of Viṣṇu in the form of a serpent. He is one of the *avatārapuruṣa*s mentioned in verse 21.

(31) **A vulture, which has only little strength, rests in the sky holding a snake in its beak for a *prahara*. Why can [the incarnation of Viṣṇu] in the form of a tortoise, who possesses an inconceivable potency, not hold the earth in the sky for a *kalpa*?**

A *prahara* is a period of time equal to about 3 hours. The description given of the vulture fits the serpent eagle better than the vulture, which does not eat snakes.[148] The tortoise referred to is the incarnation of Viṣṇu in the form of a tortoise, one of the *avatārapuruṣa*s of verse 21.

(32) **The statement that the earth has a power of attraction is not [correct], since a dense object reaches the**

[147] See *Siddhāntaśiromaṇi*, *golādhyāya*, *bhuvanakośa*, 4–5.

[148] See Minkowski 2004a, 363–364, n. 50.

earth fast, but a light object becomes attracted fast [?]. How can the earth be unmovable without support?

While the verse is not clear, the idea seems to be that a lighter object is more responsive to attraction, as it has less inertia.

(33) **Half of the sphere of the earth is above the Ocean of Salt Water. Meru is where the gods always stay. Indeed, it [Meru] alone has "upness." The lower station is said in the good *purāṇa* to be in the place where the demons [stay].**

The Ocean of Salt Water is located along the equator; beneath it, in the Southern Hemisphere, are a number of ring-shaped oceans of other liquids (see verse 39). The half of the sphere of the earth above the ocean of salt is the Northern Hemisphere. Jñānarāja holds that only Meru is "up," that is, that "up" is an absolute entity, namely, from the South Pole toward the North Pole. This differs from the astronomical tradition, which holds that "down" is always from one's feet toward the center of the earth. Jñānarāja is here following and reinterpreting the tradition of the *purāṇas*, according to which the demons are "below."

For the cosmology of the *purāṇas*, see Introduction, page 20.

(34) **If this is so, why do these mountains, oceans, rivers, and men that are located below not fall into space? An inanimate, heavy object that is not supported is seen to fall down.**

If "up" is absolute, then why do beings and objects on the southern atmosphere not fall "down" into space? We see that an inanimate object always falls "down" toward the ground when left without support.

(35–38) **What has been said is not to be doubted. Certain objects with qualities characterized by specific potencies can be different according to circumstances and place, just as moonstones, which are stones, melt [when**

exposed to moonlight] and the appearance of fire in sunstones, which are also stones. [Similarly,] diamonds float on water and a magnet attracts iron that is located very far from itself. Since the water in the ocean is very high like a mountain, ah! regions have multiple qualities.

On this half [of the sphere of the earth, i.e., the Northern Hemisphere], a difference in language, appearance, conduct, and ability is seen in different regions. How much more [would this not be the case] for people in the other half [i.e., on the Southern Hemisphere]? Therefore, they possess a potency called "fixity."

What is the use for us to go on at length talking in vain? The heavy, firm, and wide earth is supported in space by the Deity, incarnated for the sake of preventing motion of the earth. It is evident that he holds on everything that is located on the lower part of the earth.

The rationale given here is, put briefly, that certain things have unusual properties. People are also different in different regions, and thus Jñānarāja postulates that a peculiar feature of people in the Southern Hemisphere is that they possess "fixity," which allows them to remain on earth and not fall "down" into space.

Sunstones and moonstones are mythical stones that exhibit certain unusual qualities when exposed to the light of the sun and the moon, respectively. The former becomes fiery in sunlight, the latter melts in moonlight.

(39) **The Ocean of Milk [lies] after the Ocean of Salt Water. The Ocean of Ghee [lies after the Ocean of] Yoghurt [which lies after the Ocean of Milk]. [Then comes] the Ocean of Sugarcane Juice and [then] the Ocean of Liquor. After that [lies] the Fresh Water Ocean.**

According to the geography given in the *purāṇas*, there are a total of seven ring-shaped oceans filled with various liquids.[149] The order of these annular oceans presented here by Jñānarāja

[149] See Pingree 1978a, 554.

follows the order given in the *Siddhāntaśiromaṇi*.[150] However, the account differs from that found in the *purāṇas*.

(40) **In the center of the Ocean of Fresh Water lies Vaḍavāgni. From that [place], which is submerged into a large body of water, columns of smoke arise, which are carried in every direction in the sky. They dissolve when scorched by the rays of the sun, becoming the sparks of lightning.**

On the South Pole of the earth, in the center of the ocean of fresh water, lies Vaḍavāgni, a submarine fire, which causes smoke to appear out of the water. Vaḍavāgni is also known as Vaḍavānala (see verses 22–23).

(41) **Jambūdvīpa is on the top part of the earth north of the Ocean of Salt Water, which is lying [around it] like a girdle. Thus in the southern half [of the earth], there are six islands located each between two oceans.**

Jambūdvīpa, the region of the lands known to human beings, is on the Northern Hemisphere. In the Southern Hemisphere are six islands, each between two of the ring-shaped oceans. These islands are thus also ring-shaped.

(42) **[The six islands, in order, are:] Śākadvīpa, Śālmaladvīpa, Kauśadvīpa, Krauñcadvīpa, Gomedadvīpa, and Puṣkaradvīpa. The regions inside of Jambūdvīpa are called *varṣas*, or they are indicated by it [Jambūdvīpa].**

The *varṣas* are the different regions of Jambūdvīpa.

(43) **The city of Laṅkā is in the ocean [the Ocean of Salt Water], which extends 130 *yojanas*. To the east of it at [the distance of] a quarter of [the circumference of] the earth is Yamakoṭi. [East] of that [city at the same distance] is Siddhapura. And [east] of that [at the same distance] is Romakapura.**

[150] *Siddhāntaśiromaṇi*, *golādhyāya*, *bhuvanakośa*, 22–23.

The Ocean of Salt Water extends 65 *yojanas* on each side of the equator. The four cities mentioned earlier, i.e., Laṅkā, Romakapura, Siddhapura, and Yamakoṭi, are located in the middle of it. For these imagined cities, see verses 22–23 above.

(44–49) **Vaḍavāgni indeed is south of them and Meru is to the north of them. Such indeed are the six regions.**

North of the city of Laṅkā is [the] Himālaya [mountain]. Then [the] Hemakūṭa [mountain], and then [the] Niṣadha [mountain].

Likewise, [north] from Siddhapura is [the] Śṛṅgavat [mountain], [the] Śukla [mountain], and [the] Sunīla [mountain, in that order]. [These mountains] meet the ocean [the Ocean of Salt Water] in the east and the west by their length [i.e., the lengthwise ends meet the ocean as described]. Between them [that is, the mountains] is the location of the [different] *varṣas*.

Bhāratavarṣa is located in the region between Laṅkā and [the] Himālaya [mountain range]. [Then, north] from that [city of Laṅkā,] reaching up to [the] Hemakūṭa [mountain range] is Kinnaravarṣa. [The region] extending up to [the] Niṣadha [mountain range] is given as Harivarṣa by the wise.

Similarly, north of Siddhapura are Kuruvarṣa and Hiraṇmayavarṣa. [And north] of them is Ramyakavarṣa. Thus six [*varṣas* are described]. I will now explain the three western [*varṣas*].

The Mālyavat mountain is north of Yamakoṭi, while the Gandhamādana mountain is [north] from Romakapura. The two of them meet [the] Sunīla and Niṣadha [mountains]. [The region] between these four mountains is called Ilāvṛtavarṣa, the ground of which is beautiful, being covered with jewels and gold.

Meru, which resembles a lotus, is in its [i.e., Ilāvṛtavarṣa's] center, surrounded by mountain ranges on all sides.

Similar geographies are described in the *Siddhāntaśiromaṇi* and in the *purāṇas*.

Some of these regions can be identified as actual regions in the area of India. For example, the Himālaya mountains are a well-known mountain range. Other regions and mountain ranges, however, cannot easily be identified. The region that is known as Bhāratavarṣa is identified with India.

(50–51) **[The mountain Meru, which is] the abode of the gods is made up of gold and jewels; it pierces the earth, appearing [above the surface] in both ends [i.e., at the North Pole and at the South Pole].**

At its peak the gods sport. At its bottom the calm multitude of demons [dwell].

The three peaks on the mountain of the gods are multi-colored, full of shimmering gold and jewels, and the three cities [are located there]. In these [cities] Śiva, Viṣṇu, and Brahmā [dwell] always. Beneath them are the eight cities of the divine rulers of the directions.

The two verses contain a description of the Meru mountain. The statement that Meru pierces the earth through the poles means that he sees the mountain as a central axis of the universe.

(52) **It is said that Bhadrāśvavarṣa extends from the city of Yamakoṭi up to the Mālyavat mountain. Ketumālavarṣa indeed extends from Romakapura to [the] Gandhamādana [mountain]. However, this is not how it is explained by those who are learned in the *purāṇa*s.**

The geographical description given in the *purāṇa*s (for example, the *Bhāgavatapurāṇa*) is indeed different from the one given here by Jñānarāja, who is following the account of Bhāskara II in the *Siddhāntaśiromaṇi*.

(53) **The Mandara mountain and the Sugandha mountain are lying east and south, respectively, from the abode of the gods [i.e., the mountain Meru]. The Supārśva and Vipula mountains are north and west, respectively, from the abode of the gods.**

Since Meru is at the North Pole, it does not make sense to talk about something being east, north, or west of Meru.

(54) **The good banners on the peaks of [these four] mountains are, in order, a *kadamba* tree, a *jambū* tree, a *pippala* tree [the sacred fig tree], and a *vaṭa* tree. Vaibhrajaka, Dhṛti, Nandana, and Caitraratha are the groves to be known, respectively.**

The modern botanical names of the trees mentioned are *adina cordifolia* for the *kadamba* tree, *syzygium cumini* for the *jambū* tree,[151] *ficus religiosa* for the *pippala* tree, and *ficus indica* for the *vaṭa* tree. The latter is commonly known as the banyan tree.

(55–56) **In the groves on the mountains are lakes: the Aruṇa, and after that the Mānasa, and the Mahāhrada, and Sitajala. [Sitajala] is auspicious by virtue of its lotuses, and it is crowded with a multitude of geese playing on its waters.**

The delightfully charming young women in the groves that are the abode of the god of love, [women] whose very dark eyes are like a multitude of restless bees [i.e., their glances are like restless bees] and whose faces are lovely and round like golden lotuses, stay in the water with the gods.

A poetic description of the beauty found in the mountain groves is given.

(57) **The Jambūnadī [River] springs from the streams of fluid flowing from the *jambū* fruits there. Mixed with soil, it becomes the gold known as Jāmbūnada. This island [i.e., Jambūdvīpa] is named after the pleasant abode of these celebrated trees.**

When the fruits from the *jambū* trees there fall to the ground, they break and subsequently release their juice, which forms a river. The liquid of this river is then mixed with the soil,

[151] While the *jambū* tree, after which Jambūdvīpa is named, is almost always identified with the rose apple tree (*syzygium jambos*), Wujastyk has shown that it actually is a plum tree (*syzygium cumini*) (see Wujastyk 2004).

and a type of gold known as Jāmbūnada is formed from this combination.

(58) **The [heavenly] Gaṅgā fell from the sky onto [the] Meru [mountain], flowed into the lakes on the peak [of Meru] high upon the supporting mountains, [and] flowed into Bhadrāśvavarṣa, Ketumālavarṣa, Kuruvarṣa, and Bhāratavarṣa; it gives liberation even in *kaliyuga* to those who submerge themselves in it.**

The stream of the heavenly Gaṅgā flows into the four regions mentioned after landing on the peak of Meru. The stream that enters Bhāratavarṣa becomes the river Ganges in India, a river considered sacred in Hinduism.

For the *kaliyuga*, see commentary on *Siddhāntasundara* 2.1.2–3.

(59) **The interior of Bhāratavarṣa is said [to have nine] divisions, [namely,] Aindra, Kaśeru, Tāmraparṇa, Gabhasti, Kumārika, Saumya, Nāga, Vāruṇa, and Gāndharva.**

(60) **The [seven] mountains [i.e., mountain ranges] in Bhāratavarṣa are [called] Māhendra, Śukti, Malaya, Ṛkṣaka, Pāriyātra, Sahya, and Vindhya. [In this way] all the divisions on the surface of the earth with their great mountains, towns, forests, lakes, and so on are described. Afterward [the interior of the earth] from the *pātāla*s is narrated.**

The Sahya and Vindhya mountain ranges are actual mountain ranges in India. The identification of the others, however, is not straightforward. The seven *pātāla*s are subterranean worlds, the names of which will be given in the next two verses.

(61–62) **In the hemisphere of the earth, in the interior, are seven hollow spaces; they are called *pātāla*s. In these, the [inhabitants of the] world of serpents see owing to the sun[-bright] light of the gems on the hoods**

of the great serpents.

The seven *pātāla*s are [as follows]. Atala, Vitala, and the one beginning with *ni* [i.e., Nitala]. Below these is another called Gabhastimat. The next two are the ones beginning with *mahā* and *su*, [respectively, that is, Mahātala and Sutala]. [The last one is] Pātāla [or Pātālatala].

The worlds in the interior of the earth, the *pātāla*s, are considered to be inhabited by serpents. The serpents have gems on their hoods, and these gems provide the light that enables the inhabitants to see. Seven *pātāla*s are mentioned.

(63[a–b]) **In the *pātāla*s there are the [colors] black, white, red, yellow, gravelly, stony, and golden.**

Each of the seven *pātāla*s have an associated color, listed here.

(63[c–d]) **The serpent [Śeṣa], who on the underside supports the earth, who is resting on the tortoise, who stepped with his foot across the earth's surface, sometimes [becomes] one with his head bent down by the weight of the earth, and then there is an earthquake. This is the view of the *saṃhitā*s.**

Śeṣa, one of the *avatārapuruṣa*s mentioned in verse 21, supports the earth from below. When the weight of the earth causes his head to move, earthquakes occur. The *saṃhitā*s mentioned here are treatises on divination of various kinds; they are the authority for the present theory of earthquakes.

(64) **[The wind called] Bhūvāyu abides 12 *yojana*s from [the surface of] the earth. The clouds [exist] in it. After that is [the wind] called Āvaha. After that is the wind called Pravaha, which has a westward motion. After that are the Udvaha and Saṃvaha [winds]. Two other winds are the Parivaha and the Parāvaha.**

The multitude of stars along with the planets move amidst these [winds pushed] by the Pravaha [wind].

This verse describes the seven cosmic winds. In the Indian astronomical tradition, it is considered that these winds are the cause of planetary motion, moving the planets and the stars.[152]

In the *Siddhāntaśiromaṇi*, Bhūvāyu and Āvaha are considered the same wind, and Suvaha is inserted between Saṃvaha and Parivaha.[153]

(65) **The two polestars are in the sky at the top and bottom of Meru. The circle of stars is rotating, being between the polestars. As if situated like a piece of iron between two stones named loadstones, which are in the sky, the circle of stars does not fall down.**

There are two polestars in the Indian tradition. One is directly above the earth's North Pole, whereas the other is directly below the South Pole. Since India is on the Northern Hemisphere, the former is the most important of the two polestars.

When the fixed stars move across the heavens, the polestars stay fixed. The image given here is that the polestars stabilize the motion of the stars, preventing them from falling down.

(66–68) **The Creator placed the disk of the moon, which is an ornament in the form of a sphere of water, at 51,566 [*yojanas*] from the center of the earth; Mercury at 166,032 [*yojanas* from the center of the earth]; Venus, which consists of light, at 424,088 [*yojanas* from the center of the earth]; the sun at 689,377; Mars at 1,296,619 [*yojanas* from the center of the earth]; Jupiter at 8,176,538 [*yojanas* from the center of the earth]; Saturn at 20,319,071 [*yojanas* from the center of the earth]; and the circle of the stars, which is evenly marked with invisible constellations beginning with Aśvinī and bound to the pair of polestars, in the sky far from [the orbits of] all [the planets] at 41,362,658 [*yojanas* from the center of the earth].**

As noted in the Introduction (see page 14), planetary mo-

[152] See Introduction, page 14.

[153] *Siddhāntaśiromaṇi*, *golādhyāya*, *madhyagativāsanādhikāra*, 1.

Planet	Geocentric distance in *yojanas*
The moon	51,566
Mercury	166,032
Venus	424,088
The sun	689,377
Mars	1,296,619
Jupiter	8,176,538
Saturn	20,319,071
The fixed stars	41,362,658

Table 1: Geocentric distances of the planets

tion is described using epicycles in Indian astronomy. As such, the planets do not move on a perfect circle around the earth. However, each planet has a mean distance to the earth, and it is these distances that are given here by Jñānarāja. The geocentric distances of the planets as presented here are shown in Figure 1.

The values of the geocentric distances given in the three verses come from the assumptions that the circumference of the moon's orbit is exactly 324,000 *yojanas* and that each planet traverses the same amount of *yojanas* over the same period of time. Using the value $\sqrt{10}$ for π, the geocentric distances can be found.

Indian treatises on astronomy generally give the circumferences of each planet's orbit, but Jñānarāja chose to give the radii of the orbit instead.

(69–75) **The demonstration by Bhāskara [II], Pṛthūdakasvāmin, and others is not given here in our *tantra*. I am presenting a [demonstration] that is effective in counteracting the opinions of adversaries, is the opinion of noble-minded people, and is very pleasing.**

The time between the rising and the setting of the moon on the local horizon is one's own "day"; it is established by means of a water clock. Whatever is [arrived at] in this case [as the *ghaṭikās*] in the "day" from the motion of the moon by means of the method in the Section on Three Questions [on Diurnal Motion], that is the measure of the "day" for a person in the center of

the earth.

If *yojanas* corresponding to the radius of the earth multiplied by two times [the motion of] the moon are [arrived at] by means of the difference in *ghaṭikās* of the two [times], then what is [arrived at] by means of the *ghaṭikās* of one's own "day and night" [i.e., the period between one moonrise and the next]? By means of a proportion, the answer is the orbit in *yojanas* of the moon. Alternatively, this is to be computed from a syzygy with the sun.

If this orbit of the moon is [arrived at] by means of the true velocity [of the moon], then what [is arrived at] by means of the mean velocity? In this case, [the answer is] the mean [orbit]. The product of that and the revolutions of the moon [in a *kalpa*] is the orbit of heaven. That [orbit of heaven] divided by the revolutions of a planet [in a *kalpa*] is [the planet's] own orbit.

The orbits [of the planets] multiplied by the radius and divided by the minutes of arc in the degrees of a revolution [i.e., 21,600 minutes of arc] are the geocentric distances in *yojanas*.

The diameter [of a circle] is approximately the square root of [the result of] the division of the square of the circumference [of the circle] by 10. The accurate [value is found as follows]. The [trigonometric] radius multiplied by two [is the divisor when the 21,600 minutes of arc] in the circle of stars [is the dividend], [all of which is multiplied by the circumference].

The total motion of a planet in a [*mahā*]*yuga* [that is, in 4,320,000 years] is 18,712,080,864,000 *yojanas*. The earth has a circumference of 5,059 [*yojanas*].

Thus the geocentric distances in *yojanas*, known from a good demonstration, are given, as well as the measure of the circumference of the earth agreed to by [both] demonstration and the *āgamas*.

For the explanation of these verses, consider Figure 1. The small circle in the center is the earth and the larger circle the apparent path of the moon around the earth in its daily rotation. The center of the earth is O, and the given location is P. The

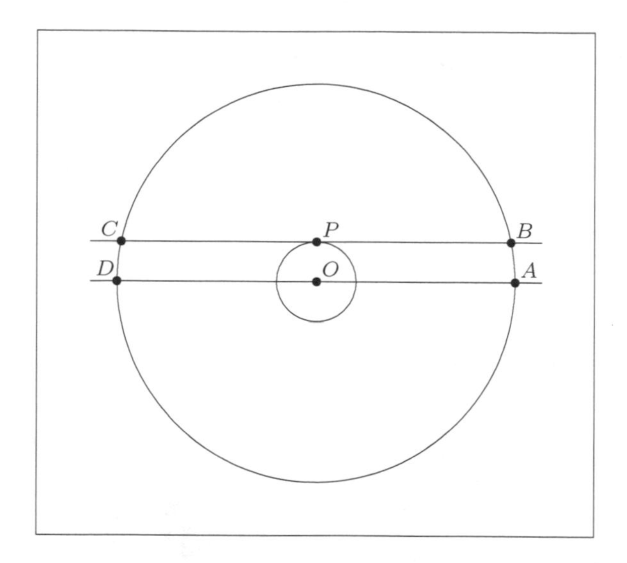

Figure 1: The geocentric distance of the moon

line CPB is the horizon at P, and the line DOA is a parallel line through the center of the earth, at which point Jñanarāja imagines an "observer."

For a person at P, the moon rises when it is at the point B and sets when it is at the point C. Let us assume that this takes t *ghaṭikā*s (for the unit *ghaṭikā*, see the Introduction, page 16). Now, if we compute the times for the rising and the setting of the moon according to the methods given in this work, because the earth is considered a point, the times will be valid for Jñānarāja's imagined "observer" at the center of the earth, but not for an observer displaced from the center at P. For Jñānarāja's imagined "observer," the moon "rises" when it is at the point A and "sets" when it is at the point D. Clearly the time it takes for the moon to travel from A to D is greater than t *ghaṭikā*s. Let us say it is $t + \tau$ *ghaṭikā*s.

Now, the extra distance traveled by the moon from A to D is the arc AB and the arc CD. Each of these arcs is roughly equal to the distance OP, which is the radius of the earth, as can be seen on the figure. Hence, the two arcs are together roughly equal to two earth radii. Let m be the circumference of circle $ABCD$ and σ the duration from one rising of the moon to the next. Jñānarāja uses a simple proportion to relate the time τ for traversing the two arcs and the time σ for traversing the whole circle as follows:

$$\frac{\tau}{2 \cdot |OP|} = \frac{\sigma}{m}, \tag{5}$$

and thus

$$m = \frac{2 \cdot |OP| \cdot \sigma}{\tau}. \tag{6}$$

The circle $ABCD$ is equal to the size of the moon's true orbit at the given time, because the moon is located at the distance of its orbital radius. The time between two successive observed moonrises determines the true velocity of the moon, whereas using the moon's mean velocity to calculate σ and τ gives us a value for m that represents the size of the mean orbit of the moon.

The product of the moon's orbit and the revolutions of the moon in a *kalpa* is the orbit of heaven. This is because the cosmology of the *siddhānta*s assumes that each planet travels the same number of *yojana*s in a *kalpa*, namely, 18,712,080,864,000 *yojana*s,[154] which is also the orbit of heaven.[155] So, by dividing the orbit of heaven by the revolutions of a given planet, we can find the orbit of that planet.

To find the geocentric distance g of the planet, we note that g is the orbital radius of a circle of circumference c. Then, since $2 \cdot \pi \approx \frac{21{,}600}{3{,}438}$,[156] we have that

$$g \approx \frac{c}{\frac{21{,}600}{3{,}438}} = \frac{3{,}438 \cdot c}{21{,}600}. \tag{7}$$

Jñānarāja gives two methods for computing the diameter d of a circle from its circumference c. The first is equivalent to the expression

$$d = \sqrt{\frac{c^2}{10}}, \tag{8}$$

which is an approximation and based on $\pi \approx \sqrt{10}$. The second, which he uses to compute g, is equivalent to the expression

$$d = \frac{3{,}438 \cdot c}{21{,}600}, \tag{9}$$

[154] The same value of the orbit of heaven is given in *Sūryasiddhānta* 12.90.

[155] See Pingree 1978a, 556.

[156] This will be discussed in greater detail in the section on true motion, the beginning of which deals with trigonometry; see *Siddhāntasundara* 2.2.2–14.

which is based on $2 \cdot \pi \approx \frac{21,600}{3,438}$. Jñānarāja says that the latter method is exact, but it is in fact also approximate, because it also uses an approximate value of π. Taking $\pi = \sqrt{10}$ in cosmological computations is common; the approximation is also found in the mathematics of the Jain community of India.

The value given for the circumference of the earth is 5,059 *yojanas*, but see verses 25–26 above.

(76) **The shapes, measures, and motions of the earth, planets, and stars given by the followers of the *purāṇas*, which are ultimately true, are indeed for another *kalpa*. Now, [in this present *kalpa*,] the [shapes, and so on of the earth, and so on] given in the treatises that give knowledge of time [i.e., *jyotiḥśāstra*] are to be thoroughly studied by the wise.**

The contradictions between the cosmology of the *purāṇas* and the cosmology of the astronomical tradition are here explained as an instance of *kalpabheda* (see verse 19 above and the commentary thereon).

(77) **He who knows the variegated body of the Cosmic Being that comprises everything and is spoken about by the ancient sages attains intimate union with the Supreme Being; thus is the meaning of the words of Vedavyāsa, [the compiler of the *purāṇas*].**

Knowledge of the cosmos is here presented as a means to spiritual liberation. This verse echoes passages in the *purāṇas*.

(78) **Whose body-encircler [i.e., the lower part of a *śivaliṅga*] is the earth with its oceans, whose ultimate [upward] staff [i.e., the upper, phallic part of a *śivaliṅga*] is [the mountain] Meru, whose base is the tortoise [i.e., one of the *avatārapuruṣas* supporting the earth], for its bath the clouds which are moving abodes of water, and for its fruits and flowers for worship the stars, moon, and planets, and for its waving of the sacred lamp the sun, may that *jyotirliṅga* worshipped by Brahmā be even within me.**

A *śivaliṅga* (called *jyotirliṅga* in the verse) is a phallic symbol used to worship the god Śiva. In the verse, the world is portrayed as a *śivaliṅga* worshipped by the creator-god Brahmā. Note that the verse does not describe the tusk of the elephant-headed god Gaṇeśa.

(79) **[Thus] the form of the universe in the *golādhyāya* is explained in the beautiful and abundant *tantra* composed by Jñānarāja, the son of Nāganātha, which is the foundation of [any] library.**

2. Chapter on Cosmology

Section 2: *Rationale of Planetary Motion*

This section of the *Siddhāntasundara* gives the theory behind planetary motion, including cosmic winds moving heavenly bodies.

(1) **At noon, according to the solar day at the beginning of the bright *pakṣa* in the month of *madhu*, the Creator [that is, the god Brahmā], residing in Siddhapura, attached the circle of the stars and the planets, who were at the beginning of [the zodiacal sign] Aries, to the Pravaha wind, which blows in a westward direction. Thereupon the courses of the year and so on, and all the motions of the planets simultaneously began; they did not [commence] at the beginning of the *kalpa*.**

The description is not entirely clear. Why, for example, is Brahmā, the god of creation, said to be in Siddhapura? At any rate, the main point is clear, namely, that Brahmā attaches the stars and the planets to one of the cosmic winds, called Pravaha, which carries the fixed stars across the heavens.[157]

That the motion of the heavenly bodies did not commence at the beginning of the *kalpa* follows from the *saurapakṣa*'s idea of a *śṛṣṭikāla*, a period of creation during which the planets do not move.[158]

(2–3) **"Tied to the wind, [the planets], which possess their own forms and are headed by the sun, and the stars always circumambulate from left to right [the deity] Parameśvara of three forms [that is, those of Viṣṇu,**

[157]The cosmic winds were described by Jñānarāja in the previous section (*Siddhāntasundara* 1.1.64); see also Introduction, page 14.

[158]See *Siddhāntasundara* 1.1.20 and accompanying commentary.

Śiva, and Brahmā], who is on the Meru mountain."—If [such an argument is made], [we counter that it is] not to be seriously considered, [because if so], how is human suffering caused by [the planets] moving through the zodiacal signs? They [that is, the planets] always instruct human beings for the good. The fruition of past actions inevitably comes about.

Jñānarāja here presents an argument, which he subsequently refutes. The argument, plainly stated, asserts that the motion of the planets and the stars is mechanical. The cosmic winds carry the heavenly bodies around the Meru mountain, the home of the god Parameśvara,Parameśvara the combined form of the three gods Viṣṇu, Śiva, and Brahmā. The motion is such that the heavenly bodies circumambulate the deity from left to right, as is considered auspicious in the Indian tradition.

However, if the motion of the heavenly bodies is merely mechanical, how can their positions among the zodiacal signs influence human beings? In other words, a purely mechanical model would invalidate astrology, which Jñānarāja believes in, and which in medieval India was closely connected with astronomy. Since the motion of the planets and their location on the zodiac instruct mankind, Jñānarāja rejects the argument.

(4) **Dense *māyā* covers this world like a shadow. Within [the world] is the tree of the sphere of celestial bodies, which yields good fruits, and which has the polestar as its root. The planets headed by the sun, which move about in the constellation-branches [of the tree] that [in turn] move due to the wind, always bestow good and bad fruits [of past actions] on human beings according to their destiny.**

In this verse, Jñānarāja envisions the cosmos as an enormous tree. The zodiacal constellations represent the branches of the tree. These branches move as a result of the cosmic wind, and the planets move along the branches with their own independent motion. They bestow both good and bad on human beings.

The Sanskrit term *māyā* means "illusion." It is the illusion by which humans fail to see the world as it is and how it works.

(5) **For the inhabitants of Siddhapura, the planets began to move eastwardly from the zenith. They are being led by the winds, their apogees, declinations, and nodes, as well as [moving by] their own force.**

Siddhapura is one of the four mythical cities (described in *Siddhāntasundara* 1.1.22–23) on the terrestrial equator. Why the commencement of the planets' motion is here described with reference to the inhabitants of Siddhapura is not entirely clear, but the idea is that the motion commenced at their zenith, that is, right above Siddhapura.

The cosmic winds are mentioned as a cause of planetary motion, but other factors, which will be elaborated upon in the next verse, are introduced as well, namely, a planet's apogee, declination, and node. Finally, it is emphasized that the planets have their own agency as well when it comes to their motion.

At this point in the treatise, Jñānarāja has not defined what he means by "apogee," "declination," or "node." The problem is that in this chapter, the *golādhyāya*, which is theoretical rather than actually giving working astronomical formulae, it is basically assumed that the reader knows the terminology used by the author. Many of the terms will become clear in the next chapter, the *grahagaṇitādhyāya*. In fact, reading the *grahagaṇitādhyāya* first prepares the reader to understand the material of the *golādhyāya* better.

For the technical terms used here, the reader is referred to the discussion on Indian astronomy in the Introduction (see page 13).

(6) **There are said to be six kinds of motion: east-west motion [caused] by [the planets'] own force and the [cosmic] winds; north-south motion [caused] by [the planets' distance from their] nodes and the declinations [of the planets]; and epicyclic motion caused by [the planets'] own epicycle[s].**

Planetary motion is here divided into six different types. A planet's east-west motion is partly due to the cosmic winds, which move the fixed stars, but the planet also has its own motion, independent of the motion of the stars; this motion is attributed to the planet's own force. The motion of the planet in the north-south direction is caused by the planet moving not on a circle equal or parallel to the celestial equator, but rather on its own inclined orbit. Finally, the fact that a planet is influenced by one (in the case of the luminaries) or two (in the case of the star-planets) epicycles also affects its motion.

(7) **Just as an insect on the rim of a wheel that is kept in motion by a potter is [itself] moving in the opposite [direction] of the rotation [of the wheel], so also a planet on the circle of the stars appears to move along with the fast-moving circle [though it has its own independent motion on it].**

Jñānarāja here explains the planets' independent motion via the example of an insect on a potter's wheel. An insect on the rim of a turning wheel appears to be moving in the direction of the wheel, though it has its own independent motion on the wheel in the opposite direction. Similarly, on the one hand the planets are moved along by the steady motion of the celestial sphere, and on the other hand they have their own independent motion.

(8) **How is this potter's wheel of time perceived in the case of the westward rotation [of the celestial equator]? The *ghaṭikā*s produced by it bestow the ripening of good and bad [fruits of past actions] to human beings. The polestar is its axis and the stars all around on it are clouds [?]. The planets, like the insect, revolve on its circumference. [Such is] the revolution of the circle of circles [?].**

The example of an insect on a potter's wheel is continued here. The turning of the celestial equator is used to define various time units; as it turns, time passes. As time passes, various good and bad results of past actions happen to human beings. The remainder of the verse is not entirely clear.

(9) **Some people say that for [each of] the moon, Mer-**

cury, Venus, the sun, Mars, Jupiter, Saturn, and the stars there are spheres rotating about the earth that are transparent like crystal. According to this opinion, on account of the fixed positions of the planetary orbits, their [i.e., the planets' and stars'] rotation, [including] the perigees, apogees, nodes, and so on, is fixed by the polestar. It is not our opinion due to the weight of their forms.

This particular verse has already been discussed in the Introduction (see page 33).

(10–11) **"Due to the speed of the Pravaha wind, a planet revolves rapidly on the circle of the stars [while] facing the other direction. The planets, headed by the sun, are withdrawing from the east when they have a westward motion due to their weight."—[Such ideas] are not to be entertained. Just as a planet has motion in the north and south directions and fast, slow, and retrograde motion, having separated from the [fixed] stars, so also, it moves toward the east due to its own force.**

The argument presented here to then be refuted is not clear, nor is it made clear whose opinion it reflects. Jñānarāja again makes the point that a planet has its own force, from which part of its motion comes.

(12–13) **Just as the celestial latitude, which is explained as the sine of an arc, after having left the [*śīghra*?] apogee returns [to it], in the same way, due to the force of the wind, [it] slowly departs elsewhere from an eastern position. Likewise, when a planet located at the *śīghra* apogee approaches the perigee, then, since it has a western motion, it is said to be retrograde.**

If [it is argued that] that the planets, which deviate, as it were, from the position of the mean orbit, abandon their own eastward motion due to contact with the perigee and [then] become retrograde, then how is it that everybody sees the entire form of the planet [?] perceiving its broad disk.

The verses are not clear. A planet moving from its *śīghra* apogee to its *śīghra* perigee is not necessarily retrograde, for example.

Jñānarāja discusses the distances from the *śīghra* apogee at which a planet becomes retrograde in *Siddhāntasundara* 2.2.29–31; the reader is referred to this passage and the accompanying commentary for more information about the relationship between a planet's distance to its *śīghra* apogee and its retrograde motion.

The word *vipadasya* is here translated as "planet," which appears to be the best interpretation.

(14) **A planet, having various forms, located at its apogee with respect to the earth is near the earth and has a broad disk [visible to observers on the earth], just as [a planet] located close to or far from the sun is said to be at its perigee or its apogee [respectively].**

The phrasing is confused. If a planet is near its apogee, it is far from the earth and thus its disk is perceived as small by an observer on the earth's surface. However, if it is close to the perigee, it is close to the earth and its disk appears larger. Presumably this is the intended meaning.

(15) **The apogee is that by which a planet, [when] attracted by it, abandons its mean orbit and moves far from the earth. [Similarly,] the perigee is that by which a planet, attracted by it, is close to the earth and has a conspicuous form.**

The definition of apogee and perigee given here makes sense. As was noted in the commentary to the previous verse, a planet close to the apogee is far from the earth, whereas a planet close to its perigee is close to the earth.

(16) **The node [of a planet] is that which pushes a planet toward the pole of the ecliptic. It is invisible and has a westward motion.**

A node of a planet is the intersection between the planet's inclined orbit and the ecliptic. As such, when a planet is far from its node, it is closer to one of the poles of the ecliptic. The node, being really a theoretical construct, has no physical manifestation; its motion is to the west, the opposite direction of the motion of the planets (when they are not retrograde).

(17) **It is said that a planet, which has simultaneous motion in three directions, moves eastward due to the Pravaha wind. Its rising and setting occur at day and at night on account of the earth.**

Jñānarāja repeats that the primary cause of a planet's eastward motion is the *pravaha* wind. The planet's setting and rising are of course relative to the earth. A planet rises when it becomes visible on the horizon and sets when it moves below the horizon.

(18) **What is [said] in the *purāṇa*, [namely, that] the rising and setting of the planets is due to the mountain Meru, that appears to be true since the Meru mountain is located centrally in the earth [as an axis].**

In the cosmology of the *purāṇas*, a planet rises when it becomes visible from behind Meru and sets when it moves behind Meru. At first glance, this appears to contradict Indian astronomy, but Jñānarāja, who interprets Meru as a central axis of the universe (see *Siddhāntasundara* 1.1.50–51), argues that the two are not contradictory.

(19) **"Like an amla fruit on a thread, the sphere of the earth is at the center of Meru [conceived of as an axis]. Beyond it, in the sky, the planets move about, having a motion toward the right for the *asuras*. They always rise and set, as it were, on account of the Golden Mountain [i.e., Meru] or [it can be explained] from the circle of the earth."—such is known by the people.**

The earth is viewed as being at the center of a straight string, which is the Meru mountain, or the axis of the universe. The *asuras* are a group of powerful divine beings that are opposed to the gods, who in the *purāṇas* dwell in the nether regions of the universe. In the Indian astronomical tradition they inhabit the South Pole, whereas the gods inhabit the North Pole. For people on the Northern Hemisphere, the planets move toward the left, but if you are on the Southern Hemisphere, as are the *asuras*, they move to the right.

(20) **[The time] from [one] sunrise to [the next] sunrise is a day that is termed civil with respect to the sun, [or just] a civil day. [Similarly], a sidereal day is [the time] from [one] rising of the stars [to the next]. A *ghaṭikā* is said to be the 60th part of a [civil] day.**

Jñānarāja here defines two types of days. A civil day is the time from one sunset to the next, and a sidereal day is the time from the rising of a fixed star to the next time it rises.

(21) **A planetary day [of any planet] is greater than a sidereal day on account of the eastward motion of the planet. [The amount by which it is larger is measured] by the minutes of arc in the result from the rising time of the zodiacal sign of the planet [?] multiplied by [the planet's] velocity and divided by 1,800.**

A planetary day, or a day of a particular planet, is a full revolution of that planet; for example, the time between the planet rising and the next time it rises. The formula given is, however, not clear.

(22) **[The length of] a *saura* year is [the duration of] the sun's journey through all of the zodiacal signs. That [period of time] is [also known as] a divine day and night. The difference of the two is [a day] of the ancestors [?].**

There are two *kalpas* of [the god] Brahmā: when the sun is present it is [Brahmā's] day, and the other [*kalpa*, when the sun is not present], is [his] night. The total of the two [*kalpas*] is said to be [Brahmā's] day and night.

A year is the time it takes the sun to travel the full length of the ecliptic, here called a *saura* (literally "solar") year. More information on the *saura* year is given in *Siddhāntasundara* 2.1.2–3 and the accompanying commentary.

A year is referred to as a divine day and night because if you are located on the North Pole, as are the gods, you will see the sun when it is above the celestial equator, but not when it is below it. In other words, for half of the year the sun is visible, so it is day, and in the other half it is below the horizon, so it is night.

The definition given of a day and night of the ancestors, who live on the moon, is unclear. In the Indian tradition, a lunar month is generally taken to constitute a day and night of the ancestors.

The day and night of the creator-god Brahmā are said to last a *kalpa* each. When Brahmā sleeps, the universe is destroyed, so the sun is only present during his day (see verse 26 below).

(23–24) **The prime vertical along the terrestrial equator is said to be the horizon for those dwelling on Meru. When the sun is in the six zodiacal signs beginning with Aries, which are north of there [that is, north of the equator], it is their [that is, the dwellers on Meru] day; [when it is] elsewhere than there [that is, in the other six signs], it is [their] night. When the sun is at the beginning of [the zodiacal sign] Cancer, it is noon [for them, and] when it is at the beginning of Capricorn it is midnight [for them]; that is to be accepted by a wise man.**

According to what is said in the *purāṇa*—"[Their] day [commences] from the beginning of Capricorn"— the beginning of their day is at midnight.

As has already been noted, Jñānarāja does not always define the technical vocabulary used in the *golādhyāya* (see the discussion on Indian astronomy in the Introduction, page 13). The statement that when the sun is in one of the six signs that begin with Aries, that is, in Aries, Taurus, Gemini, Cancer, Leo, or Virgo, it is night for the gods, who dwell on Meru, only makes sense if the zodiac is tropical, not if the zodiac is sidereal. In other words, the vernal equinox coincides with the beginning point of Aries. Normally the Indian zodiac is sidereal, but in certain contexts precession is taken into account, which amounts to using a tropical zodiac.

(25) **Noon for the ancestors, who dwell on the surface of the moon, occurs at the end of the conjunction of the sun and the moon, and [their] midnight is at the end of the full moon [i.e., at the end of the opposition of the sun and the moon]. Sunrise and sunset occur for them, [respectively,] in the two halves of the lunar month known as the dark and the bright [*pakṣa*].**

A day and night of the ancestors are here considered to be a lunar month.

(26) **[When he is] awake, Brahmā beholds everything, beginning with the disk of the sun, up until the end of the *kalpa*, [then] goes to sleep. During [his] sleep, everything is dissolved. Therefore, there are two *kalpa*s, said to be his day and night.**

In verse 22 it was already stated that both the day and the night of Brahmā last for a *kalpa*. More information is given here, namely, that the universe is destroyed during Brahmā's night, when he is asleep.

(27) **A solar month is [measured] by 30 of our own days and nights. It is the same as the days in the passage of the sun from one zodiacal sign to the next. Similarly, a lunar [month] is a day and a night of the ancestors. A year is equal to 12 months.**

Definitions of solar and lunar months are given here. A solar month is the time that it takes for the sun to traverse 30° of the zodiac, that is, a zodiacal sign. A year is defined to be 12 solar months, or the time that it takes for the sun to traverse the entire zodiac.

(28) **At the time when day and night of [Brahmā], the Creator, are equal [in duration], there is a complete destruction of the world [?]. Therefore, the proper com-**

putation of the motion of the planets by means of [their] daily motion was made by the wise.

The first half of the verse appears to be corrupted, and the meaning is unclear. In fact, Brahmā's day and night are always of the same duration, each a *kalpa*.

(29) **The eastward motion of a planet is of two kinds, true and mean. The true [motion] of a planet, which stays on its eccentric orbit, is [furthermore] of three kinds according to the division into fast, slow, and retrograde [motion].**

The information here is very brief, and it is unreasonable to expand upon it here. In fact, mean motion is the topic of the first section of the *grahagaṇitādhyāya*, whereas true motion is the topic of the second. More information is given there.

(30) **Since the true motion is not the same every day due to the pulls of the perigee, apogee, and the wind, therefore the mean velocity of a planet is derived from the day count by means of a proportion.**

The information is again very brief. Basically, the velocity of a planet is not constant, but changes over time. A mean velocity can be defined. More information on this is available in the first and second sections of the *grahagaṇitādhyāya*.

(31) **[Thus] ends [the section on] the cause of [planetary] motion in the *golādhyāya* in the beautiful and abundant *tantra* composed by Jñānarāja, the son of Nāganātha, which is the foundation of [any] library.**

3. Chapter on Cosmology

Section 3: *Method of Projections*

This section covers the theoretical framework of the Indian planetary model.

(1) **He who supports the earth, on the surface of which there are gods, mountains, and clouds, and who is causing the planets to move in the wind on the revolving circle of stars for the sake of the good of all people and born of his own play—may that blessed Hari [i.e., Viṣṇu], the support of everything, fulfill my most precious desire.**

Jñānarāja opens this section with a verse in praise of the god Viṣṇu. Viṣṇu is described as the one who supports the earth and causes the movement of the heavenly bodies. He does so for the good of all living beings and as part of his playfulness.

(2–3) **The circular orbit [of a planet, that is, the concentric circle], which resembles the ecliptic, is produced from the mean geocentric distance from the center of the earth and is given in *yojana*s.**

The circle, on which the planets, by their mean velocity, complete the [number of] revolutions declared [for them] in a *kalpa*, and which is located entirely upward from the [circular] orbit by the *yojana*s in the sine of the greatest equation, is the circle that is named [after having an] apogee [i.e., it is the eccentric circle]. At the bottom of that [circle] by the [same distance in *yojana*s from the concentric circle] is what is known as the perigee.

[There are two such eccentric circles, namely,] the *manda*, due to the slow motion [of the planet], and the *śīghra*, due to the fast [motion of the planet]. The cir-

cle centered at the mean planet with [the radius determined] by [the *yojanas* of the sine of the greatest equation of] the *manda* circle is called the *manda* epicycle.

These two verses give the basics of the model used in Indian astronomy. The universe is considered to have a spherical earth in its center, around which the heavenly bodies revolve. Since a planet does not move in a perfect circle around the earth, its distance to the earth changes with time. However, each planet has a mean distance to the earth, and the circle centered at the earth and with this mean distance as its radius is called the *concentric circle* of the planet. A theoretical construct, the *mean planet*, moves on the concentric circle.

In reality the planet moves on a different circle, called the *eccentric circle*. As such, the *true planet* moves on the eccentric circle. The eccentric circle has the same radius as the concentric circle, but its center is displaced from the position of the earth by a certain distance. It is on this circle that each planet completes a certain number of revolutions per *kalpa*, a number that is given in the astronomical treatises.

The eccentric circle has two special points on it, namely, the point farthest from the earth, which is called the *apogee*, and the point closest to the earth, which is called the *perigee*. The apogee and the perigee are located directly opposite each other on the eccentric circle. In the Sanskrit texts on astronomy, the eccentric circle is literally called the "circle with an apogee."

For the two luminaries, the sun and the moon, there is only one eccentric circle, but for the five star-planets, there are two. One of them is called *manda* ("slow") and the other *śīghra* ("fast"). These will be discussed in great detail in *Siddhāntasundara* 2.2.15 and the accompanying commentary.

The "equation" is the angular distance between the mean position of the planet and its true position on the epicycle. In Figure 3 below, the equation is $\angle SEM$. The "greatest equation" is the largest angular distance between the mean planet and the true planet; its sine is here said to equal the distance between the centers of the concentric and eccentric circles.

The concentric and eccentric circles are shown in Figure 2. The point E represents the earth, and the circle centered at E is the concentric circle. The point E' is the center of the eccentric

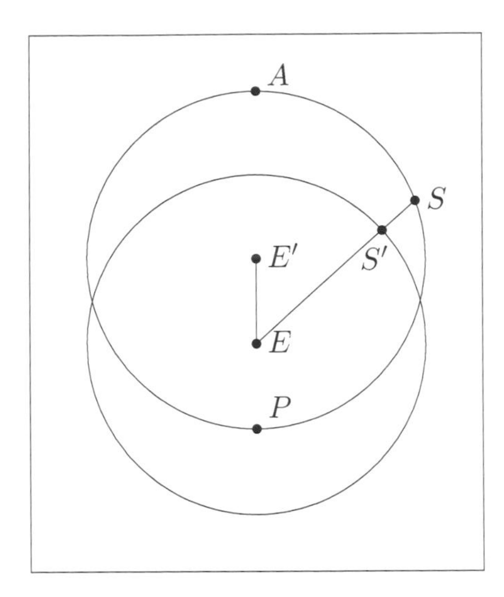

Figure 2: The concentric and eccentric circles

circle, on which the planet moves. The point S represents the planet on the eccentric circle; its position, seen from earth, is given with reference to the concentric circle, that is, the point S'. The apogee and the perigee on the eccentric circle are marked by points A and P.

(4) **[If] the [mean] geocentric distance of a planet, [measured] in *yojanas*, is [derived] by means of the radius [of the reference circle], then what is [derived] by means of the sine of the greatest equation? That which is found are the *yojanas* in the sine of the greatest equation, respectively for the *manda* and *śīghra* circles.**

The verse gives a simple proportion for computing the size of the greatest equation of a planet measured in *yojanas*. Let R be the radius of the reference circle[159] and $\text{Sin}(\mu)$ be the sine[160] of the greatest equation of the planet, and further let D be the mean geocentric distance measured in *yojanas* and d the

[159] In the *Siddhāntasundara*, R is generally 3,438; see the discussion in the section on true motion in the *grahagaṇitādhyāya*.

[160] Sines are discussed in the beginning of the section on true motion in the *grahagaṇitādhyāya*.

greatest equation also measured in *yojanas*. Then the following proportion holds:

$$\frac{\text{Sin}(\mu)}{R} = \frac{d}{D}. \tag{10}$$

In other words, we have that

$$d = \frac{D \cdot \text{Sin}(\mu)}{R}. \tag{11}$$

From this equation, we can find the greatest equation of the planet measured in *yojanas*, if we know the mean geocentric distance of the planet in *yojanas* and the sine of the greatest equation with respect to the reference circle. Since this equals the displacement of the center of the eccentric circle from the position of the earth, we have therefore found the distance between the concentric and eccentric circles.

(5) **The manda-corrected [planet] never moves below the *manda* perigee or above its apogee. Therefore, the corrected [planet is found] from the *śīghra* epicycle. The mean [planet] never departs from the orbit-circle.**

That a planet is *manda*-corrected means that its position has been corrected by means of the *manda* eccentric circle. Such a *manda*-corrected planet will inhabit a position on the *manda* eccentric circle, which in particular means that it cannot move beyond the boundaries indicated by the apogee and perigee of the circle. Since a planet does travel past these boundaries, the position found by this correction is not the true position of the planet. To find the true position of the planet, it is necessary to also correct the planet by means of the *śīghra* epicycle. It is not explained here how to combine the effects of the two epicycles, but see *Siddhāntasundara* 2.2.23^{c}–24 and the accompanying commentary.

By definition, the mean planet is always located on the concentric circle.

(6) **The *yojana*[s] of the sine of greatest equation [of a planet] is on [the respective] epicycle by geocentric distances increased by the *manda* and *śīghra* perigees of the [given] planet [?].**

The *manda* and *śīghra* epicycles of a planet always move eastward on their own circle with the velocity of their *manda* and *śīghra* motion [respectively].

The verse is not entirely clear, but the intent appears to be to mark off the sine of a planet's greatest equation measured in *yojana*s.

(7) **The eccentric circle, which resembles the circular orbit [that is, the concentric circle], [in that its radius is] produced by the geocentric distance in *yojana*s, [has its] center displaced from the center of the earth toward the apogee by the *yojana*s in the greatest equation.**

The verse merely says that the concentric and eccentric circles, which naturally have the same radius, differ in their position in that the eccentric circle is displaced from the concentric circle by the *yojana*s in the greatest equation of the planet.

(8) **Considering that on the eccentric circle the beginning point of Aries is [found] to the left of the intersection of the *manda* apogee and the *manda* eccentric circle [?] by the [number of] degrees of the *manda* apogee, it follows that the *manda*-corrected [planet on the eccentric circle] is [found] by the [number of] degrees of the mean planet from the [beginning point of Aries on the eccentric circle] in regular motion.**

As can be seen in Figure 3, the position of the *manda*-corrected planet occupies the intersection of the *manda* epicycle and the eccentric circle, at the point S (though there may be two points of intersection depending on the position of the mean planet). Furthermore, the point E represents the earth, the center of the concentric circle, and E' the center of the eccentric circle. M is the mean planet. Note that, naturally, the distances EE' and MS are equal.

What is meant in the verse is clearly that the point of intersection occupied by the *manda*-corrected planet corresponds to the similar point on the concentric circle, as is clear from Figure 3. As such, the verse simply states that positions on the ec-

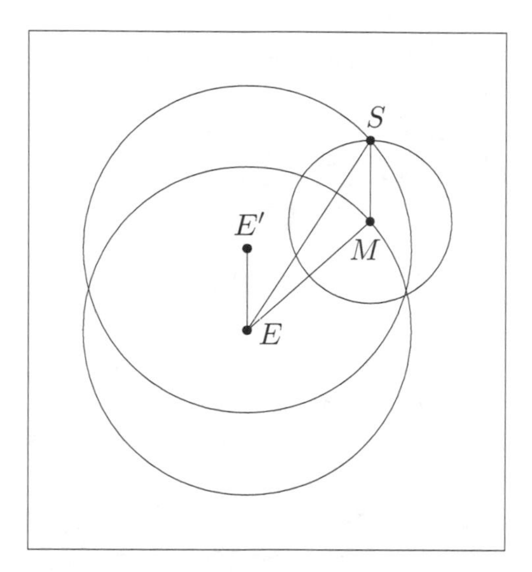

Figure 3: The concentric circle, the eccentric circle, and the epicycle

centric circle are equivalent to positions on the concentric circle. Even though specific reference is given to the *manda* eccentric and concentric circles, the same argument would hold for the *śīghra* eccentric and concentric circles.

(9) **The string between the [position of the mean planet on the concentric circle] and the center of the earth is the mean geocentric distance [of the planet]. The position of the *manda*-corrected planet is located [on the concentric circle] where the mean orbit is intersected by the string.**

In order for the first statement of the verse to be true, the word *tat* (literally, "this" or "that") of the verse must refer to the mean planet, not the *manda*-corrected planet, as it has been translated above. However, since the word refers back to the previous verse, it more likely refers to the *manda*-corrected planet, though the distance between the center of the earth and the position of the *manda*-corrected planet is generally not the mean geocentric distance of the planet.

However, the second statement can only be correct if the

string extends from the center of the earth to the position of the *manda*-corrected planet, in which case the position of the *manda*-corrected planet with respect to the concentric circle, that is, the mean orbit, is given by the intersection of this circle and the string.

(10) **The *manda* equation is the [angular] difference between [the longitudes of] the [*manda*-corrected planet] and the mean [planet]. It is to be found by means of the science of computation. The anomaly is the [angular] difference between the *manda* apogee and the mean [planet]. The center of the epicycle is the mean [planet].**

As seen in Figure 3, the *manda* equation is the angle SEM between the longitude of the *manda*-corrected planet and the longitude of the mean planet. Similarly, the anomaly is the angle $E'EM$ between the *manda* apogee and the mean planet.

(11–13) **The apogee-string is the string stretched from the center of the earth to the apogee point [on the eccentric circle]. In like manner, having determined the intersection of the string touching the center of the mean orbit toward the geocentric orbit [i.e, the eccentric circle] [?], the sine of the greatest equation is always the distance between the intersection of the strings and that [?].**

The eccentric-circle-string is the leg or the upright located between the apogee of the planet and the string.

When the anomaly is in [the six signs beginning with] Capricorn or in [the six signs beginning with] Cancer, the upright is [respectively] above or below its own sine of the greatest equation. Therefore, their [?] sum or difference is here the upright and leg; the hypotenuse and the equation are [found] from the leg.

The two verses aim to provide a geometric illustration of the greatest equation of a planet, but the construction is too unclear to be reconstructed precisely. The definition of the eccentric-circle-string is not clear either.

(14) **Since the planet proceeds in front of the *manda* apogee with its fast [*śīghra*] motion, [it follows] that [the mean planet's longitude] diminished by [the longitude of] the *manda* apogee, called the anomaly, is the difference between [the longitudes of] the planet and its apogee.**

Since the planet is [moving] backward from the *śīghra* apogee with its slow [*manda*] motion, [it follows] that [the [longitude of] the *śīghra* [apogee] diminished by [the longitude of] the [mean] planet is the *śīghra* anomaly; the *śīghra* equation is said [to be derived] by means of the [*śīghra* anomaly].

This lengthy verse essentially does nothing more than define the anomalies belonging to the *manda* and *śīghra* epicycles, respectively, as the angular difference between the apogee and the longitude of the mean planet.

(15) **The *śīghra* epicycle, which is all around the planet corrected by the *manda* equation and located on the mean orbit by the *yojanas*, is, like the *manda* [epicycle], to be imagined having the marks of the degrees of a circle at a location far from earth.**

Jñānarāja here addresses the nature of the *śīghra* epicycle. It is clear from the description that the *śīghra* epicycle acts on the planet after it has been corrected by the *manda* epicycle, and just as the *manda* epicycle, it is envisioned as a circle with its center on the mean orbit, that is, the concentric circle.

(16) **The apogee and perigee are 180 degrees crosswise [from each other via] a string through the center [i.e., the line between the apogee and perigee passes through the center of the earth].**

The beginning of [the zodiacal sign] Aries is [at a location] from the apogee determined by the degrees of the apogee. The [position of the] planet, in order, is conceived to be [measured with reference to] the beginning of Aries.

It is clear that the apogee and the perigee are separated by exactly 180 degrees on the eccentric circle and that the line joining them passes through the center of the earth. Similarly, by the definition of the apogee, the angular distance between it and the beginning point of Aries is precisely the longitude of the apogee. The position of a planet is measured with respect to the beginning of Aries.

(17) **When the anomaly is at the beginning of [the zodiacal sign] Capricorn, it [the true planet] is above the radius [i.e., it is outside of the concentric circle]; when it is at the beginning of [the zodiacal sign] Scorpio, it is below [i.e., it is inside the concentric circle]. The equation in this case is called "the upright." Therefore, the sum or the difference of them [i.e., the longitudes of the mean planet and the apogee] is called "upright"; [it and] the leg corresponding to it [form a right triangle with] the *śīghra* hypotenuse [i.e., the radius of the *śīghra* epicycle].**

The description of the anomaly being at the beginning of Capricorn means that the anomaly is 90 degrees, and similarly it is 270 degrees when it is at the beginning of Scorpio.

The situation corresponds to the planet and its apogee being either 90 degrees or 270 degrees apart.

(18) **If the *bhujaphala* is at the tip of the hypotenuse, what measure of the arc is at the tip of the radius? Whatever is the arc of the result, that is [called] the *śīghraphala*, which is added to or subtracted from [the longitude of] the *manda*-corrected [planet].**

When the tip is at the side, from the *manda*-corrected planet, in the true planet, from the anomaly[?]; it is to be known, the method for the daily motion of the planet.

The verse explains how to combine the contributions from the two epicycles. The geometrical model utilized here is discussed in detail in the section on true motion in the *grahagaṇitā-*

dhyāya; in particular, the *bhujaphala* is defined and discussed in *Siddhāntasundara* 2.2.19 and the commentary thereon. The last half of the verse appears to be corrupted and is not clear at all.

(19) **The difference between [the longitude of] a planet tomorrow and [its longitude] today is said to be the [daily] motion of the planet, divided into [the contributions of] the *manda* and the *śīghra* [epicycles].**

For the sake of computing this [daily motion], the motion of the center of the planet is found via a proportion with the true [planet] and with the divisor and multiplicand being the motions.

The difference between the position of the planet tomorrow and its position today naturally gives the distance it traversed during a day, that is, its daily motion. The proportion in the second half of the verse is not clear.

(20–21) **Wherever the moon or the sun is on the concentric circle, the center of the *śīghra* epicycle of [the planets] beginning with Mars is at that position [?].**

If the proportion involving the difference of the sines divided by 225 minutes of arc does not give the [right result?], then what is attained by the motion of the anomaly? And what [is attained] by 360 degrees on the *manda* epicycle?

The first part of the verse does not, as it stands, make complete sense. Clearly the position of the moon on the concentric circle is not connected with the center of the *śīghra* epicycle of any star-planet. The verse continues to introduce a proportion, which also does not make complete sense. It is not specified which "sine difference" we are talking about, for example.

(22) **[Thus] ends [the section on] the method of projections in the beautiful and abundant *tantra* composed by Jñānarāja, the son of Nāganātha, which is the foundation of [any] library.**

4. Chapter on Cosmology

Section 4: *Description of the Great Circles*

This section of the *Siddhāntasundara* is dedicated to a description of the great circles that play a role in Indian astronomy.

(1–2) **The east-west [circle] that passes through the local zenith is called the prime vertical, and the north-south [circle passing through the zenith] is called the meridian. The [circle] known as the horizon for those that dwell at the center of the earth is at a distance of 90 degrees from each of their [i.e., the prime vertical and the meridian] intersections. The local horizon [for a person on the surface of the earth] is elevated from that [horizon just described] on all sides by the number of *yojanas* in the radius of the earth. Here [i.e., on the local horizon] the reddish-colored disk of the sun is seen in an enclosure made level by means of water.**

In the opening verses of the section, Jñānarāja defines three basic circles in the Indian cosmological model. The local horizon is the great circle that marks the boundary between the part of the sky that is visible and the part that is invisible as a result of being obscured by the earth, the prime vertical is the great circle that passes through the zenith and intersects the local horizon at its east and west points, and the meridian is the great circle that passes through the zenith and the North and South Poles. These have all been discussed in the Introduction (see page 16), where also the circles to be defined in the next verses are mentioned.

(3) **[The circle that is] to the south of the zenith by the degrees in the [local] latitude, to the north of the *pātālas*, and fixed at the east and west points is known as the equator.**

The next circle to be defined is the celestial equator, which is the great circle derived from projecting the terrestrial equator onto the sphere of the stars.

(4) **The circle that passes through the two polestars and the circumference of which passes through the east and west points is said to be the horizon in the city of Laṅkā. Elsewhere it is [called] the six-o'clock circle.**

The six o'clock circle is the circle with the property that when the sun first reaches it on a given day, it is 6 in the morning, and when it reaches it a second time, it is 6 in the evening. At the city of Laṅkā, which is on the terrestrial equator, sunrise and sunset always occur at 6 in the morning and 6 in the evening, respectively, which means that the local horizon at Laṅkā is precisely the six o'clock circle.

(5) **By whichever path it [that is, the sun] traverses all the visible constellations with an eastward velocity, by that [path] the form of the band of the circle of constellations [that is, the ecliptic] is [seen] in the sky at a given location.**

This verse describes the ecliptic, that is, the path of the sun across the heavens, though its interpretation and meaning are not entirely clear. The idea is that the sun follows a particular path—a great circle—in the sky, which is called the ecliptic. It is along this great circle that the constellations of the zodiac, Aries and so on, are found.

(6–7) **The beginning point of [the zodiacal sign] Cancer is 24 degrees north of the equator; the beginning point of [the zodiacal sign] Capricorn is [the same number of degrees] south [of the equator].**

[The longitude of a planet] diminished by [the longitude of] the node [can be found] from the declination [of the planet].

The two endpoints of [the zodiacal signs] Pisces and Virgo are attached to it [i.e., the equator].

Elsewhere the zodiacal signs are either south or north

of the equator by the degrees of their respective declinations. They, along with the ecliptic, are moving with the wind, and they never depart from their individual boundaries.

The celestial equator and the ecliptic do not coincide. The angle by which the ecliptic is displaced from the celestial equator is called the obliquity of the ecliptic, denoted by ε; in the Indian tradition $\varepsilon = 24°$.

In the following, we need to assume a tropical zodiac. There are two intersections of the equator and the ecliptic: at the point where the sun crosses the equator from south to north is called the vernal equinox, and the other is called the autumnal equinox. In a tropical zodiac, the autumnal equinox corresponds to the beginning point of the zodiacal sign Aries (or the endpoint of Pisces). The beginning point of Cancer is 90° away, so it corresponds to the point where the ecliptic is farthest north of the equator. Similarly, the beginning point of Capricorn corresponds to the point where the ecliptic is the farthest south of the equator.

In the formula given, "node" must refer to the vernal equinox. If we know the planet's declination, that is, the angular distance between it and the equator, we can find its angular distance from the equinox, which we will refer to as the planet's precessional longitude; the formula is explained in detail for the sun in *Siddhāntasundara* 2.2.32$^{\text{c–d}}$ and the accompanying commentary.

(8) **The inclined orbit of a planet is attached to the ecliptic at the [two] node[s], [each node being] six zodiacal signs [that is, 180 degrees] from the [other] node.**

[A planet's] own greatest celestial latitude always [occurs] between the circumferences of the two circles [that is, the ecliptic and the planet's inclined orbit] three signs from [either of] the two nodes.

Just like the ecliptic, the path of the sun, each planet has an orbit, which is a great circle attached to the ecliptic at two points, called nodes. The two nodes are opposite each other on this great circle, 180° apart. When a planet is exactly 90° from one of its nodes, it is the farthest from the ecliptic as it can get, either to the north or to the south, depending on which node it is 90° from.

(9) **When a planet is at [one of] its node[s], there is no celestial latitude; when [it] is a distance of three zodiacal signs from [one of] its node[s], the greatest celestial latitude [occurs].**

The computation of the celestial latitude is said by the ancients to be by means of the sine of the arc [corresponding to the longitude] of the planet diminished by the [longitude of] its node via the rule *madhye 'nupāta*, etc.

The rule for computing the latitude of a planet works precisely as the one for the declination given in verses 6–7 above. The source or full details of the abbreviated rule mentioned are not clear.

(10) **The degrees of the declination [of a planet] are the degrees, the terrestrial latitude, and so on by which the position of the planet on the ecliptic is [removed] from the equator. The [planet's] celestial latitude, on the other hand, is the distance between the two positions [i.e., its true position and its longitude on the ecliptic] of the disk of the planet.**

Bringing in the terrestrial latitude as a component of the declination seems awkward here, but see *Siddhāntasundara* 2.1.27–29[a].

(11–12) **The south and north [poles] of the equator are said by the great sages to be at the two polestars, [whereas] the south and north [poles] of the ecliptic are located at a distance of 24 degrees from the [corresponding] polestars. Owing to the rotation of the circle of the stars, [the poles of the ecliptic] rotate around the polestars. It [that is, the ecliptic pole] is known as *kadamba*.**

The *ayanavalana* is the distance between the [ecliptic pole] and the horizon.

The Sanskrit word *kadamba* is the name of a tree (nuclei cadamba), but the word is also used in astronomy to denote the pole of the ecliptic. Since the obliquity of the ecliptic is $\varepsilon = 24°$, it is clear that the poles of the eclectic are 24° from the polestars.

As is so often the case in the *golādhyāya*, Jñānarāja uses technical terms with either no or a very brief definition: in this case, the term *ayanavalana*. The *valana*s, one of which is the *ayanavalana*, are used to compute eclipses. For a full and detailed discussion, see *Siddhāntasundara* 2.5.22–32 and the accompanying commentary.

(13) **The diurnal circle [of a planet] is at all places to the south or north of the equator by the degrees of its corrected declination. That [circle] is to be known as the "day-and-night circle." The measure [i.e., radius] of the diurnal circle is half its diameter.**

The diurnal circle, here called the "day-and-night circle," and sometimes in English called the "day circle," is the circle described by a planet during a 24-hour period when it revolves with the fixed stars, and during which its independent motion is very slight. The last statement is obvious.

(14) **[The circle known as] the *dṛṅmaṇḍala* has a circumference that is attached to the disk of the [given] planet and passes through the zenith of the observer.**

For the sake [of computing] the latitudinal parallax and the longitudinal parallax, this [circle] is [called] the *dṛkkṣepamaṇḍala*, if it is attached to the nonagesimal.

Two great circles are defined here, namely, the *dṛṅmaṇḍala* and the *dṛkkṣepamaṇḍala*. The former passes through the position of a given planet and the zenith of the observer, whereas the latter is a secondary circle to the ecliptic which passes through the zenith of the observer. Note in particular that the *dṛkkṣepamaṇḍala* must pass through the nonagesimal.

The word *kṣitigarbhagarbham* in the second quarter of the verse is problematic and not translated in the above.

(15–16) **Since the ecliptic and the orbit of a planet are in motion, the *dṛṅmaṇḍala* is also moving.**

The circle for fixing the position of a planet is the one that moves while fixed on pins at the two ecliptic poles.

The pair of prime verticals is immobile, [as is] the north-south circle [the meridian?] and the triad of horizons [?]. The diurnal circle is moving.

The construction of the circles is done by means of an armillary sphere.

The two verses make a distinction between moving and nonmoving circles. The ecliptic and planetary orbits are examples of the former, whereas the prime vertical, the meridian, and the horizon are immobile. It is not clear what Jñānarāja has in mind when he refers to the "pair of prime verticals" and the "triad of horizons"; the two verses do not go into full detail and neglect to discuss the other circles, such as the six o'clock circle, the *dṛṅmaṇḍala*, and so on.

The distinction between moving and nonmoving circles is not one that will play a role later.

(17) **[Thus] ends [the section on] the description of the great circles in the *golādhyāya* in the beautiful and abundant *tantra* composed by Jñānarāja, the son of Nāganātha, which is the foundation of [any] library.**

5. Chapter on Cosmology

Section 5: *Astronomical Instruments*

This section discusses astronomical instruments, which are necessary for the actual practice of astronomy.

(1) **Since a *tantra* possesses astonishment through [the use of] instruments, therefore I will here explain [some of these] instruments, [such as] the *cakra-yantra*, the *turya-yantra*, the *sambhramagola-yantra*, the *stambha-yantra*, the *māyūra-yantra*, the *ghaṭī-yantra*, and the *saikata-yantra*.**

Jñānarāja here praises the use of astronomical instruments (*yantra*), through which the astonishment found in an astronomical treatise is brought to life. In the following, a description of a number of astronomical instruments will be given.[161]

(2–3) **The *turya-yantra* [that is, the sine quadrant] is to be constructed so as to have the form of a quarter circle. A pair of sighting vanes with holes are to be placed on it. An index arm and plumb line are [fastened] on it in a hole at the apex. On the circumference 90 degrees are to be beautifully marked. Thirty equidistant [lines of] Sines are [marked] on it [that is, the sine quadrant] beneath the sighting vanes extending perpendicularly down to the degree [markings]. One *ghaṭikā* is to be marked by means of 6 degrees [on the circumference]. The index arm is marked with *aṅgula*s corresponding to the space between the [lines of] sines.**

Verse 3 and the following verses deal with the *turya-yantra*,

[161] For a detailed overview of astronomical instruments in India, see Ohashi 1994.

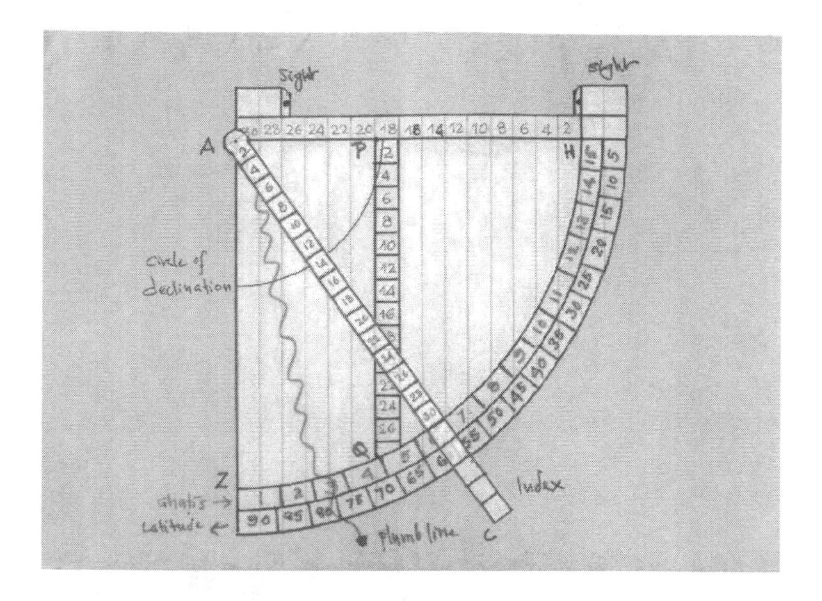

Figure 4: Sine quadrant

that is, the sine quadrant. This instrument was first mentioned in India in the *Brāhmasphuṭasiddhānta* of Brahmagupta.[162] A careful exposition of the sine quadrant in India has been given by Sarma.[163]

The instrument is shaped as a quarter circle and shown in Figure 4.[164] A pair of sighting vanes are found on top of the instrument, and an index arm and a plumb line are attached at the apex. The arc is marked twice, once with 90°, and then in 15 equally spaced parts, 6° each, corresponding to 15 *ghaṭikās*. In addition, 30 equidistant lines, called lines of Sines, are marked on the instrument beneath the sighting vanes. Finally, the index arm is divided into equal pieces corresponding in size to the space between the equidistant lines on the instrument; the 30th mark on the arm will touch the rim of the instrument, before the markings for *ghaṭikās* and degrees.

Owing to the 30 equidistant lines, we take the radius of the quarter circle to be $R = 30$.

[162]See Sarma 2008, 2.

[163]See Sarma 2008. Relevant to this topic are also Sarma 2012 and Sarma 2011.

[164]The figure is a drawing by S. R. Sarma, used with permission.

(4) **The radius [of the quarter circle] diminished by [respectively] the versed Sine of one [degree], two [degrees], three [degrees], and so on is [one] leg [of a right triangle] and the radius is the hypotenuse. The square root of [the difference of] the squares of these is the given leg. In this way the [lines of] sines are to be placed on the sine quadrant.**

The construction proceeds through a number of right triangles. Let $R = 30$ denote the radius of the quarter circle. By the definition of the versed Sine ($R - R \cdot \text{vers}(n) = R \cdot \cos(n)$ for $n = 1, 2, 3, \ldots, 89$), the first triangle has legs Cos(1°) and Sin(1°); the second Cos(2°) and Sin(2°); the third Cos(3°) and Sin(3°); and so on up to Cos(89°) and Sin(89°). The hypotenuse of each triangle is, of course, R. The trouble is that this does not produce 30 equidistant lines on the instrument, and so it does not produce the lines of sines mentioned in verses 3–4.

(5–6) **The zenith is defined [as the point] beneath the apex. Crosswise from that [apex] is the horizon. Rising times [?], signs, degrees, and minutes of arc are [to be marked] on the top of the instrument. A portion is to be marked with *pala*s and so on.**

The sine of the obliquity of the ecliptic is to be marked on the index arm. From the apex the index arm is to be placed on the tip of the degrees on the arc. The degrees on the arc are the degrees of declination at the tip of the sine touching the sine of the obliquity of the ecliptic.

The points Z and H on Figure 4 mark the zenith and the horizon, respectively.

With the radius $R = 30$, we get $R \cdot \sin(24°) = 12;12$. This point is marked on the index arm and used to describe a circle.

(7) **When the index arm is at the degrees on the arc corresponding to [the longitude of] the moon increased by [the longitude of] its node [and] when the half *aṅgula*s corresponding to the degrees [of the longitude] of the**

sun are marked from the center, the degrees of the latitude, at the tip of the sine of the upright, multiplied by 20, at a time diminished by 19, then it is the *parvan* of the moon [that is, conjunction or opposition of the sun and the moon, or quadrature of the moon].

The verse aims at using the instruments for finding the *parvan*, which refers to one of four phases of the moon: conjunction of the sun and the moon, opposition of the sun and the moon, and the two quadratures of the moon. However, the contents of the verse are not entirely clear.

(8) **At the end of the conjunction [of the sun and the moon], the index arm located at the degrees corresponding to the difference between [the longitude of] the sun and the rising point of the ecliptic at noon is attached to the mark of the declination. The *ghaṭikā*s [indicated] at the tip of the upright are the colatitude. The degree of the zenith distance is at the rising of the sky [?] and the latitudinal parallax is half the sine of that [?].**

(9) **At noon, the inversed degrees of the inclination of the declination are the measure of the degrees in *pala*s at the tip. The equinoctial shadow from the sine of 18° up to the base touches the index arm.**

(10) **The equinoctial increased by its 60th part is to be indicated on the index arm joined to the degrees [corresponding to] the arc of the sun. The measure of *ghaṭikā*s in the motion up to the sky is from the tip of the straight sine located at that mark.**

(11) **It is to be placed down on the index arm known as the sine of the ascensional difference located at the tip of the degrees of the declination from the sky. The earth sine is the difference between touching it [?]. The earth sine is the sine situated at its sine of the declination**

from the center.

(12) **As [the longitude of] the sun is the difference between what is above the center and the hypotenuse, so one's own vision is the difference between the radius and the hypotenuse. Having fixed and beheld the sun, a planet, or a star, the degrees of the hour angle are on the circumference attached to the tip of the colatitude from the horizon.**

(13) **The sine of the hour angle is corrected by the index arm located at the degrees of the zenith distance from the middle [of the sky] or by the sine measured by its mark. The arc multiplied by** 90 **and divided by** 6 **is the *ghaṭikā*s from a proportion at noon in Laṅkā [or] in one's own city.**

(14) **Having imagined the hypotenuse above the center, its gnomon is the sum of or the difference between the shadow and the hypotenuse when the perpendicular of seven *aṅgula*s or the *aṅgula*s measured by** 12 **is located in the sky.**

(15) **The rising *ghaṭikā*s are multiplied by** 2 **toward the degree in a sign from the beginning and end of the *pala* [time]. The precession-corrected rising point of the ecliptic is [given by] the given *ghaṭikā*s of the precession-corrected sun. The lapsed *ghaṭikā*s since midday are the zenith distance; from midnight it is the complement of the zenith distance. Whoever is in the sky or in the *pātāla*s at [sun]rise in Laṅkā is at [what is] fixed on the *ghaṭikā*s of the zenith distance and complement of the zenith distance.**

As has been noted, the Indian zodiac is sidereal. However, in some contexts, it is necessary to take precession into account.

In this case, it is the rising point of a tropical zodiac that is intended with “precession-corrected rising point.”

(16–17) **The sine measured by the firm length at the mark at the tip of a tree, wherever it is attached to the mark of the *aṅgula*s [?] by means of an upright, [it is then] moreover taken to the mark of the *aṅgula*s diminished by 1.**

The distance between that and the ground to the east is multiplied by 12. The measure of the tree is increased by the visible height when the upright is located 8 degrees on the circle [?]. The measure of the earth diminished by the circle increased by visible height is said to be the height.

(18) **Having placed the center of the planet’s *manda* circle here, the 12th part of the circumference of the *manda* circle touches it at the tip of the index arm. The result is the multiplier [?] from the tip.**

(19) **Half of [the longitude of] the sun is [attained] from the multiplication or division of the *ghaṭikā*s up to the hypotenuse [?] by the sine of the degrees of the zenith distance. The day is elevated or depressed [?]—ah!—the *ghaṭikā*s are the elevation of the mark for the sky [?]**

Having made the circle, the circumference, and the base, accompanied by the perpendicular, rotate from the center like a fan, having observed the sun, one should compute the time as previously explained.

(20) **Thus the garland of astronomical instruments is stated in the beautiful and abundant *tantra* composed by Jñānarāja, the son of Nāganātha, which is the foundation of [any] library.**

6. Chapter on Cosmology

Section 6: *Description of the Seasons*

Being a poetic description of the seasons, this section of the *Siddhāntasundara* truly stands out from the remainder of the work. The poem has already been described in the Introduction (see page 35).

The existence of two registers of meaning for many verses in the *ṛtuvarṇana* poem means that it is necessary to provide two translations for such verses. The two translations are separated by a forward slash (that is, "/") below. Occasionally, the meanings of the two registers are similar enough to warrant only one translation. In these cases, however, there will be one or more words that carry a double meaning, which will therefore be translated twice, the two translations again separated by a forward slash.

As was noted in the Introduction, the following translation is a slightly revised version of that done by Minkowski and Knudsen.[165]

Many of the verses draw on stories from the great Indian epics, the *Mahābhārata* and the *Rāmāyaṇa*.[166]

(1) **The view of the astronomers of ancient times was that when the sky was obscured with clouds, one could determine the position of the sun in the zodiac judging by the signs of deep winter and the other seasons. Therefore, the spring and other seasons are described here.**

It is not clear which ancient astronomer Jñānarāja has in mind, but most likely Varāhamihira is among them, since he

[165] See Minkowski and Knudsen 2011.

[166] The editions used are Sukthankar et al. 1933–66 and Bhatt et al. 1960–75.

begins with deep winter when presenting an order of the seasons.[167] Jñānarāja, however, follows the annual order of the seasons as given by Bhāskara II in his *ṛtuvarṇana* and begins with spring.[168]

Jñānarāja's deviation from the order of the seasons used by the ancients is most likely due to an adherence to the order used by Bhāskara II and to the fact that the section on spring is much longer and contains more elaborate verses than the section on cold winter, and as such ought to have a more prominent position in the poem. It is also possible that the order reflects a deference to the god Kṛṣṇa, the god whose activities are also described in the verses on spring.

(2) **When sweet fortunate spring arrives, the Mādhavī creeper is triumphant in her forest grove. Spring greets and delights goodly people with the music that is born from the bamboos as they are filled with the wind [he] brings up.**

/

When sweet fortunate Kṛṣṇa comes, his beloved exults in the forest dwelling. Kṛṣṇa delights and greets goodly people with the music born from the flute that he fills with his breath.

(3) **Spring is dressed in yellow clothes in the form of flowers, and adorned in his heart by the sun. Spring's head is made beautiful and honored with the flowers of the Mimosa, Trumpet-flower, Campaka, and Jasmine.**

/

Kṛṣṇa is decorated with flowers and [dressed in] saffron yellow clothes. Kṛṣṇa is adorned on his chest with the Syamantaka jewel. Kṛṣṇa's head is made beautiful and honored with the flowers of the Mimosa, Trumpet-flower, Campaka, and Jasmine.

---

[167] See, for example, *Bṛhatsaṃhitā* 3.23–24 and 30.22.

[168] Kālidāsa begins the *Ṛtusaṃhāra* with summer.

The Syamantaka was the gift of the sun to Kṛṣṇa, hence *dyumaṇi*, literally "day-jewel."

(4) **[Spring's] best friend [the sun] displays a powerful brilliance, his rays elevated just from seeing him. [The sun], in fulfillment of his own desire, moves northward.**

/

Upon seeing him, [Kṛṣṇa's] best friend [Arjuna] has a beaming face. He raises his hand [to wave in greeting]. In fulfillment of his own desire he moves toward the place for the drinking party.

The sun is here depicted as the best friend of spring. The astronomical features of the annual motion of the sun and its increasing brightness inform the spring register. On drinking parties for Kṛṣṇa and Arjuna in spring, see, for example, *Mahābhārata* 1.214.[169]

The god Kṛṣṇa served as the prince Arjuna's chariot driver during the great battle of the *Mahābhārata* epic.

(5) **The Mango creeper, as if doing honor to the Spring, touches the earth with her hands in the form of lovely fronds and fruits. For this she is rewarded with perfumes and many gorgeous garments.**

/

[Kṛṣṇa's beloved, who is like] a mango creeper, doing honor to the lord of her fruitful season, as it were, touches the earth with hands that are as delicate as fronds and fruits. For this she is rewarded with perfumes and many gorgeous garments.

(6) **[Spring / Kṛṣṇa] enters the honey grove as his own palace, with the blessed singing of birds as its Brahmins, with the fronds that are lined with red parrots' beaks as its gateways, with forest floors as its courtyards, floors**

[169] As mentioned above, the edition used is Sukthankar et al. 1933–66.

made auspicious by aspersions in the form of nectar dripping from flowers.

Both registers take place in a grove in the forest, which is metaphorically a palace in the ways described.

(7) **Approaching [this honey-grove / palace], which has a canopy made of mangoes in fruit, and which has standards made of sprouting trees gently waving, [Spring / Kṛṣṇa] as beautiful as Love himself, seeing his love-companion with her head bowed over his feet,**

Meaning similar enough again not to require entirely separate translation. The sentence continues in the next verse.

(8) **speaks to her: "Dear one, how have you passed the year's accumulation of days in my absence?" "The rays of the moon, the brother of the *kālakūṭa* poison." "Why then the serving and remembering of my feet / roving?"**

The sentence is continued from the previous verse. The meaning of the verse is not entirely clear. Here Kṛṣṇa and Spring are depicted speaking to a cowherd girl[170] and the spring creeper, respectively. The two are similar enough to require no double translation here. It appears that the beloved, in expressing her reproach of the missing Kṛṣṇa / Spring, finds in separation even the moon's rays to be like poison.

In Indian mythology, the *kālakūṭa* poison arose from the churning of the ocean.

(9) **[She replies to him:] "Your moonlike face has as its best attribute the characteristic mark of gazelle-like eyes, and there is nectar that stirs desire in your lower lip. [Yet] when you are separated from your friend, we see a decrease in the brightness [of this face]. When joined with him, then there is an increase in brightness. The moon [on the other hand] would be combust by this**

[170] In Indian mythology, many stories are given of Kṛṣṇa's play with the cowherd girls during his youth.

excess. It is when separation from the [female] beloved burns the moon that it appears diminished."

The reply is one of reproach or jealousy by the cowherd girl or the creeper for the friend of Spring or Kṛṣṇa, who has been mentioned in verse 4. This is the sun, or in the other register, Arjuna. An irony is intended here, that there is an overturning of astronomical behavior: the moon when separated from the sun should become brighter, but this moon-face separated from its friend loses its luster. Closeness to the sun should result in diminished lunar brightness, but the opposite has happened for the friends.

(10) **[He speaks to her:] "Kṛṣṇa and Arjuna [as the light and dark] in your eye [are warriors] with bows in the form of your eyebrows, and arrows in the form of your sidelong glances. They as if reach Karṇa [as your ear] to conquer him.**

Then Karṇa, pierced, is made to fall on the ground. / Your ear, pierced, is brought down onto the palm of your hand.

Karṇa frequents the toddy grove, which is filled with flowers and roving bees, and populated by wild oxen and lordly elephants. How then can you be afraid of the moon / my face and other things?"

The response of the lover is also to be understood as juxtaposing two levels, that of the beauty of her face and that of epic characters. Her beauty is itself devastating in effect and can protect her by winning a *Mahābhārata*-like war. The significance of the toddy grove with its flowers and elephants on the level of the face is not clear, however.

Karṇa was an opponent of Kṛṣṇa and Arjuna on the battlefield; the word literally means "ear."

(11) **[She to him:] "Lord of my life, it's only your two eyes reflected in my eyes that are this dark and light, this Kṛṣṇa and Arjuna. But when [you] my life are turned away, where then are they?" Hearing this clever speech, he embraced her who was as dear to him**

as life. Between Kṛṣṇa and his beloved / Spring and his beloved there arose the greatest bliss. Why describe it?

The verse indicates that the preceding verses have been filled with wordplay between the speakers, not just wordplay by the poet about the speakers.

(12) **[He to her:] "[Your] hair is the dark, flawless [night] sky, [in which] your string of pearls are the stars. The gem in the parting of your hair is the sun, and in your lower lip is found the crescent moon. [Because the sun and moon are together in this sky], under their sway both the night lotuses and the day lilies stay open day and night in the animating [light].**

"Reflecting thus my soul, together with my heart, has set out [for you], dear one. / Reflecting thus the goose has not set out for the Mānasa lake, dear one, together with me."

There is wordplay in the terms *haṃsa* and *mānasa*. On one level they refer to the migration of the bar-headed geese over the Himālaya mountains to the Mānasa lake at the onset of the rainy season, a common feature of the poetic convention about the seasons. We are, however, supposed to be hearing about springtime. The sense appears to be that there is no need of Mānasa, when the face of the beloved creates the same conditions here.

(13) **The pool of the river Kālindī is her mass of hair, in which the line of foam is her string of pearls, on which the pair of golden geese [are her earrings?] and the two lotuses are the flowers behind her ears. Holding the serpent Kālīya that was her braid of hair / holding the serpent [Kālīya] as if it were her braid of hair, and beholding her heart that was a ball sunk in her waves, the best of men [Kṛṣṇa] embraces his beloved [as the river Yamunā] and frolics with her.**

Kṛṣṇa's amorous life, though with a different beloved, is juxtaposed with a natural feature. The juxtaposition is reversible: the beloved as the river Yamunā, the Yamunā River as beloved. The Yamunā flows clear and dark in the spring. Spring does not appear to come into either register other than incidentally.

(14) **In the horrible hot season, wicked people are tormented, frightened, and consumed by the heat of the sun. But for those who have pleased the Great Lord with acts of bathing, generosity, and worship, there is the good fortune of enjoyment.**

/

When Bhīṣma was as fierce as the hot season, the wicked [Daityas], eclipsed by the power of the wielders of bright rays [i.e., the Ādityas], undertook the practice of *tapas*. When Śiva was pleased with their bathing, generosity, and worship, they got the boon of the lordly power of [Duryodhana's] body.

This is the only verse for the hot season. It is less certain that there is a second layer of meaning. If so, then it refers to the boon from the god Śiva that the Daityas received after performing *tapas* (austerity), which brought Duryodhana into celestial embodiment to fight for them.[171] It thus connects this verse with those of the rainy season, where the second layer of meaning continues the *Mahābhārata* story line.

The Daityas are divine beings opposed to the gods (sometimes referred to as "demons"), and the Ādityas are a group of gods. Duryodhana is the adversary of Arjuna in the *Mahābhārata*.

(15) **Seeing the earth stunned and burned in the hot season, the dark rain cloud arises from the ocean as the destroyer of heat. Its lightning seems to dress it in yellow garments, as it gets nearer and nearer to the earth.**

/

[171] See *Mahābhārata* 3.240.

Seeing the innocent world burned in the heat [of oppression], [Kṛṣṇa], dark as a cloud, his yellow garments billowing, arises from the sea and comes from afar, to destroy the oppressors on earth.

For the rainy season we return to double-meaning verses, where the two registers describe the behavior of clouds and of Kṛṣṇa respectively.

(16) **[The cloud], whose origin is in heat, covers the light of the sun, moon, and stars. It makes the bamboo groves resound beautifully with its wind. Rumbling, the heavenly one rains down life-giving water here.**

/

[Kṛṣṇa] who is the source of all brilliance, drowns out [with his splendor] the light of the sun, moon, and stars. Beautifully blowing his flute, resounding, the divine one rains down ambrosia here.

(17) **The heavenly cloud, carrying showers of rain and circles of lightning, born from the [sun's] heat [on the ocean] in order to save the life as it were of the people oppressed by the horrible hot season, ended the heat and increased the strength of life.**

/

As if to save the life of his friend [Arjuna], who was pressed hard by the heat [of the military onslaught] from Bhīṣma, the charioteer [Kṛṣṇa], holding the edge of the flashing discus, relieved the pressure [on Bhīṣma], and caused righteousness to increase.

On a number of occasions Kṛṣṇa forgets his vow not to fight in the battle and has to be restrained by Arjuna from attacking Bhīṣma.[172] This verse invokes that episode and so is the most specific in the Kṛṣṇa register.

[172] See, for example, *Mahābhārata* 6.102.

(18) **Q: This one [Kṛṣṇa] approaches you silently, O *gopī*.**

A: That's a cloud.

Q: But what of his yellow garment?

A: My child, that is the lightning.

Q: What of the flawless garland on his neck?

A: That is a rainbow.

Q: How does he make the bamboo resound?

A: By wind.

Q: Are those not peacock feathers?

A: The cloud on the peak of a mountain rumbles [with thunder, and consequently] a group of peacocks appears there.

The verse is a conversation between two cowherd girls. One, the friend, attempts to warn the other that Kṛṣṇa is there to seduce her. Kṛṣṇa is again likened to a rain cloud, in a different setting. The association of peacocks with the rainy season is a well-established poetic convention.

(19) **Q: How does he block out the sun if it is Kṛṣṇa?**

A: Since he has the brilliance of a crore of suns. To save the earth, [Kṛṣṇa] has come from the ocean [in Dvārakā] at the request of the men of his side.

/

To rescue the earth [from the hot season], [the cloud] has come from the ocean, accompanied by the cries of birds.

When her friend explained it to her in this way, she entered the house, covering her head with the end of her garment. And Kṛṣṇa, pleased with that clever one, joined with her there.

The discussion continues, but the answerer of verse 18 becomes the questioner of verse 19. A second level of meaning is added in one place. The answer is itself ambiguous, and so is the gesture of covering the head and going indoors, from a pretended fear of the rain, or pretended modesty at the gaze of a strange man.

(20) **In autumn, which pleases the villagers, auspicious in its wealth of grain, filled with journeys and expeditions, and which confers the bliss of the goddess [Śāradā],**

Moonlike Rāma, who contained the ocean [with his bridge], who is the great lord of liberating wisdom, who holds the bow of Śiva in his hand, rules supreme,

/

the pleasing moon, which comprehends the ocean, which is the great lord of the stars, and which touches Śiva with its rays, shines brightly,

The sentence continues into the next verse. This section features the Rāma narrative juxtaposed with references to mythological, astronomical, and astrological features of the moon. Śiva wears the crescent moon in his hair and hence is touched by its rays. The worship of the Goddess takes place especially during nine days of the festival in this season, and Rāma's victory over Rāvaṇa is celebrated on the 10th day of that festival.

(21) **[Rāma], who took pleasure in the [golden] deer that leapt into the air, whose armament consisted only in the strung bow [of Śiva], who became the enemy of the night-rover [Rāvaṇa], who assembled a camp of many armies [outside the walls of Laṅkā], whose own allies [Vibhīṣaṇa and so on] joined him, who had Nīla and Sugrīva for allies, who had lotus-like eyes, and who was the delight of his many dependents/ministers.**

/

The moon, whose delight is the deer in the heavens, who enters into the Strung Bow [that is, Sagittarius], who is the enemy of the dark mover [that is, Rāhu], who has many camping places [that is, the *nakṣatras*], who makes conjunctions with the sun, who is the friend of Śiva, who is the auspicious opener of the night lotuses, and who delights many drinking cups.

In Indian mythology, the mark on the moon is a deer. The moon's many halting points are the *nakṣatra*s, which in the mythology are identified as the daughters of Dakṣa (a son of Brahmā) and the wives of the moon. In the mythology, Rāhu is an enemy of the sun and the moon, astronomically represented by the moon's ascending node. Śiva has a blue throat, and the moon is his hair ornament. The moon's reflection in the liquor in a drinking cup is a common theme of poets.

(22) **[Rāma], who delighted the heart of his wife when she was with him, tested her heart with fire when she had been separated from him, her beloved. That was fitting, for the sake of defending her virtue. The behavior of those who are beyond our thought is only for the purification of those who are prudent.**

/

The moon, which delights the hearts of women united with their husbands, suddenly torments with his natural beauty [the hearts of] those women separated from their beloveds. That is fitting, for the sake of protecting their virtue. The motion [of the moon] makes for a fitting penance for those haughty ones.

The Rāma register appears to justify the testing of Sītā, Rāma's wife, at the time of their reunion after the defeat of Rāvaṇa. There is another possible reading of the moon register, more specifically linked to the story of the wives of the moon, the *nakṣatras*, the daughters of Dakṣa, who protested the moon's regular absence with the result that Dakṣa inflicted the moon with consumption, whence his dark spots. If taken this way, this register would have to read that the moon burns in his own body, as a fitting penance for the improvement of haughty ones like him.

(23) **When fierce-bannered Vibhīṣaṇa had come to [Rāma's] feet for refuge, and had told him the news about [Rāma's] enemy [Rāvaṇa] and what had happened to the queen [Sītā], the king said to him, "Your conduct is that of an ally," and promised, "The position of my enemy will be given to you."**

/

When terrifying, fierce Ketu comes near the path [of the moon], he signals the behavior of the foe [Rāhu] and what will happen to the king's wife [Rohiṇī]. The lord of men says to [Ketu], "This is something other than the conduct of a friend. And the name of 'enemy' will be applied to you."

In verses 23–29, the Rāma register relates events from the *Yuddhakāṇḍa* of the *Rāmāyaṇa*, telling the story of Vibhīṣaṇa coming to Rāma, Aṅgada's embassy to Rāvaṇa's court, the death of Rāvaṇa, and the installation of Vibhīṣaṇa on the throne of Laṅkā. The reading of the moon register is more uncertain. It appears to narrate an account of the eclipse. Elements of the myth of the churning of the ocean, in which the *soma* (nectar) was created, and of the moon's abduction of Tārā, wife of Bṛhaspati, are present in the verses, but cannot be read to yield a continuous account that matches the extant versions of the legends. The "wife of the king" is problematic, as Rohiṇī, the moon's favorite wife among the daughters of Dakṣa and astronomically one of the constellations known as the *nakṣatras*, is not particularly relevant to the Rāhu legend or to eclipses. The sun is near Rohiṇī in early winter, and hence the moon is nowhere near its full brightness near Rohiṇī in autumn.

Ketu is the descending node of the moon, that is, the point where the moon crosses the path of the sun in its southward journey.

(24) **After that, at the command of the king, Aṅgada, as his own brilliance, went to [Rāma's] greatest enemy and recommended that he make peace. The messenger of Rāma spoke: "O King, whose hand is ever on the crowns, which have bright gems, of the lords of kings who are bowed down to you,**

/

After the lord's speech, the ray of the moon, which shone like a bracelet around him, went to the great enemy and was asked by him for a conjunction. The messenger of the pleasing moon spoke: "O ruling one,

who reduces the rays that fall to earth from the bright, gem-like planets, which are lords over men,

The sentence continues into the next verse. The reference to the crowns of Rāvaṇa's vassals probably alludes to Aṅgada's theft of Rāvaṇa's crown. Rāhu interferes with the astrological influences of the planets it comes near.

(25) **"Śiva bows to him, and the sun, too, is made to serve him every day. The serpent Ananta waits at his lotus-like feet in the ocean of waters. In his hand is the bow of Śiva. He gave Tārā to the monkey Sugrīva as his beloved wife, taking her from [Sugrīva's] greedy enemy [Vālin]. If you [Rāvaṇa] have taken away [Rāma's wife Sītā] by stealth, disguised as a mendicant, then you should give her back; show your respect to the Lord!"**

/

"Śiva bows his head to the moon every day [in that he wears the moon on his head] and the sun also attends upon him [every month]; at his lotus feet was Ananta, the serpent, when the ocean [of milk was churned and the moon was born]; in his planetary aspect Sagittarius becomes fierce; he gave his beloved Tārā [the wife of Bṛhaspati] to Mṛgaśiras of the beautiful neck, in reply to pleading by the enemy [Bṛhaspati, her husband] who was longing for her. If she has been taken secretly, give her back. Honor the request of the god."

Rāma is given some of the symbolism of the god Viṣṇu, of whom he is an incarnation, or a physical manifestation. For the moon, the reference to Sagittarius is a best guess. It could also refer to the crescent moon in Śiva's hair, but this has just been mentioned. The *purāṇa*s recount a story of the moon abducting Tārā from her husband, Bṛhaspati, and eventually relinquishing her.

(26) **"The multitudes [of *rākṣasas*] swiftly bow down, shattered by the power of the hands of [Rāma's] ally. Sacrifices, failing because they were blocked [by *rākṣasas*],**

are completed in his presence. Upon his ascension to power, elders, ministers, and poets rejoice. Go to him, the lord, for the greatest protection. Otherwise you will perish."

/

"Struck down by the heat of the rays of [the sun], [the moon's] ally, the zodiacal signs swiftly diminish. Ambitions, when perishing, burdened with obstacles, are quickly fulfilled in [the moon's] sight. At whose rising Jupiter, Mercury, and Venus rejoice. Take refuge in him, the celestial one. Otherwise you will perish."

Rāma register: The completed sacrifices refer to the sacrifices of Viśvāmitra previously interrupted in the *Bālakāṇḍa* and the observances of the sages in the Daṇḍaka forest described in the *Āraṇyakāṇḍa*, both of the *Rāmāyaṇa*.

Moon register: The sun's brilliance drowns out the natural astrological qualities of the signs of the zodiac, which become "combust" when he is near them in the horoscope. Jupiter, Mercury, and Venus are referred to here as the allies of moon in astrology, though actually this is not quite so. The four do constitute the set of benefic planets. The relationship of friendship is non-transitive, however. The moon is a friend to Mercury, but Mercury is neutral to the moon. Jupiter is a friend to the moon, but the moon is neutral to Jupiter. The moon, incidentally, is neutral to Rāhu and Ketu, though they are enemies to him.

(27) **"[Rāvaṇa replied:] '[Rāvaṇa] acquired a powerful body by drinking the milk of a lioness, and became immortal by drinking the *amṛta*. After that, through a blessing from the god [Śiva], that dark one conquered both gods and demons and [he acquired] new bodies. How can you, [Rāma], who are a weak, forest-dwelling ascetic, whose army consists only of fireflies, be capable of conquering him?' Say that to him, messenger. He will die in vain."**

/

"[Rāhu replies:] 'By drinking the milk of his mother, Siṃhī, he acquired a powerful body. He became immor-

tal by drinking the [stolen] *amṛta* [ambrosia]. After that, because of a blessing from the god [Brahmā?], that dark one acquired a new body, with which he conquered the drink of the gods. How are you capable of conquering him, you [moon], who are slender [crescent] and [dependent on] the ascetic [Śiva,] who dwells in forests, you whose army consists only of the stars in the sky?' Say that to him, messenger. He will die in vain."

This is a speech dictating the reply to the preceding; hence, Rāvaṇa and Rāhu speak of themselves in the third person.

(28) **Having said this, Rāvaṇa went forth, surrounded by his Brahmins and soldiers. Then there was a great battle. When in that [battle] his own army had been defeated, the Asura [i.e., Rāvaṇa] struck Sugrīva, the monkey. Seeing [Sugrīva] thus wounded with an arrow, the king [Rāma] himself struck off the head of the enemy with an arrow. Moonlike Rāma was victorious.**

/

Having said this, the fearsome one went forth, accompanied by the army of Ketu. Then there was a great battle. When his own army had been defeated, the demon struck the moon's distinctive mark, the gazelle with the beautiful neck. Seeing him thus struck with an arrow, the king [moon] himself struck off the head of the enemy with an arrow. The beautiful moon was victorious.

The moon register refers to an eclipse. Perhaps by poetic license, it is here the moon who strikes off Rāhu's head, though in the myth Viṣṇu decapitated Rāhu with the disk, not an arrow.

(29) **Then, installing Vibhīṣaṇa in the kingdom of his enemy, riding in the airborne conveyance with [Sītā] the young girl who got on board with him, and with his brothers and supporters, after returning each one to his own kingdom, he protected the worlds surrounded by elders, ministers, and poets.**

/

Putting the fearsome one in the sign of his enemy, riding in his sky-going chariot, in conjunction with the lovely young Rohiṇī, conjoining with his allies and friends, and entering into conjunction with Jupiter, Mercury, and Venus in their signs of rulership, he protected the worlds.

On the friendships of the planets see the above notes to verse 26. The moon is exalted in Taurus, near the constellation Rohiṇī.

(30) **When his elder brother, autumn, has gone away to another land, early winter cannot help but nourish the grain, his brother's child as it were, with cold breezes invited from the Himālaya mountain, as if out of brotherly affection it were keeping off the warmth of the sun.**

Autumn precedes early winter and hence is the elder brother imagined to be away on a journey. The poet's "science" is that it is the cold air that ripens the grain. Warding off the sun alludes to the parental role of keeping a child's head shaded.

(31) **In early winter, the moon [nourishes the grains], sprinkling with rays made of ambrosia the offspring that are his own, and his enemies' and friends' alike, or else the earth nourishes the grains as her own, by her liquid essence, having collected ambrosia from the lord of serpents.**

The offspring of the moon, enemies, and friends must be the plants that are associated with different planets. The lord of serpents lives underground, in the earth in the *pātāla*s, described in *Siddhāntasundara* 1.1.61–62.

(32) **When the sun is farthest away in the south, cold arrives with the advent of deep winter. Is the fire in the middle of the southern ocean able to defeat the cold sent by the strong winds from the Himālaya mountain[s]?**

As was the case with the previous pair of verses about early winter, verses 32 and 33 list the factors that are understood to cause the different seasons: the sun, the moon, the earth, and the mountains.

(33) **The "somagenic" moon destroys the heart diseases arising in people in this season, through applications of ripe forest herbs. The earth, rich in the bounty of fruits, water, and grains, sustains the lives of the people residing upon her.**

As above, the verse describes what the moon and earth do in this season to keep people alive in the coldest season. The moon as the lord of herbs develops their wealth-giving virtues.

(34) **When the sun is near the Himālaya, and the cold far away, there is excessive heat. In the hot season (*grīṣma*), even the nighttime is warm because of the rays of the sun passing above the earth's shadow. Heat comes to earth from the sun, water from the ocean, cold from the mountain, together with its own workings, in various ways, in its own time.**

The sun is "near the Himālaya" in its northernmost declinations.

(35) **Thus [ends] the description of the six seasons in the *golādhyāya* [i.e., the section on spherics] in the beautiful and abundant astronomical treatise [i.e., the *Siddhānta-sundara*] composed by Jñānarāja, the son of Nāganātha, which is the foundation of [any] library.**

7. Chapter on Mathematical Astronomy

Section 1: *Mean Motion*

The first section of the *grahagaṇitādhyāya* lays the scene for the mathematical astronomy of the *Siddhāntasundara* by discussing mean motion.

(1) **I salute Gaṇeśa, whose five lofty faces are the elephants of the quarters, in whose belly is the whole universe, whose crest-jewel is a necklace of thousands of mountains, who has the blue sky as his garment, who takes away the inner darkness, who is bearing the crescent moon, whose beauty is resplendent like tens of millions of suns, whose carrier is like a boar, and who is the greatest bestower of good.**

Jñānarāja is here praising Gaṇeśa in a verse seeking to highlight this god as the supreme god.

(2–3) **Now, [the duration of the reign of] a *manu* is [measured] by 71 [*mahā*]*yuga*s. There are 14 *manu*s in a day of Brahmā. [The duration of] 1,000 [*mahā*]*yuga*s is [measured] by [the duration of] those [14 *manu*s] increased by [the duration of] the *sandhi*s [that each are] equal to the years in a *kṛta*[*yuga*] [and are situated] at the three [types of] junctures. [Such 1,000 *mahāyuga*s constitute] a day [of Brahmā].**

A [*mahā*]*yuga* is [measured] by 4,320,000 [*saura*] years. Its four parts, the first being the *kṛta*[*yuga*], span, respectively, 432,000 *saura* years multiplied by 4, 3, 2, and 1.

As has been discussed in the Introduction (see page 16),

Indian astronomy operates with a number of world ages (*yugas*) of long duration, which are measured in *saura* years. A *saura* (literally, solar) year is what in modern terminology is called a sidereal year. It is the time that it takes for the sun to return to the same position with respect to the fixed stars when viewed from the earth. The beginning of a *saura* year occurs when the sun enters the sign Aries. It should be noted here that the zodiac in Indian astronomy is sidereal (see Introduction, page 13).

A *mahāyuga* (literally, "great world age") is a period of 4,320,000 *saura* years. It is divided into four smaller *yugas*: a *kṛtayuga* of 1,728,000 *saura* years, a *tretāyuga* of 1,296,000 *saura* years, a *dvāparayuga* of 864,000 *saura* years, and a *kaliyuga* of 432,000 *saura* years. Notice that these *yugas* are, respectively, $\frac{4}{10}$, $\frac{3}{10}$, $\frac{2}{10}$, and $\frac{1}{10}$ of the duration of a *mahāyuga*. Of these four ages, the *kṛtayuga* is considered to be a golden age. As the ages progress, they get worse (morality deteriorates, greed and other bad qualities become dominant, the life span of human beings decreases, and so on), and the *kaliyuga* is the worst of them all. According to the Indian tradition, we are currently in a *kaliyuga*.

The system of the *yugas* is not original to the Indian astronomical tradition, but has incorporated aspects from other sources. The system is found in *Mahābhārata* 12.231; *Viṣṇupurāṇa* 1.3, 3.1–3, 4.1, and 5.23; and *Manusmṛti* 1.61–86. It is furthermore an integral part of the cosmology of the *purāṇas* (see Introduction, page 20).

A *kalpa* is a day of the creator god Brahmā. It consists of 1,000 *mahāyugas*, or 4,320,000,000 *saura* years. At the end of the *kalpa*, when Brahmā's night begins, the universe is partially destroyed, and at the dawn of Brahmā's next day, there is a new creation.

A *manu* is a mythical progenitor and ruler of the earth. There are 14 *manus* during a *kalpa*, each reigning for a period of 71 *mahāyugas*. The three types of junctures (*sandhi*) mentioned in the verse are the time periods that occur before the reign of the first *manu* of the *kalpa*, between the reigns of two consecutive *manus*, and at the end of the reign of the last of them. There are thus altogether 15 such junctures, and each has the same duration as a *kṛtayuga*, that is, $\frac{4}{10}$ of a *mahāyuga*, or 1,728,000 *saura* years.

The reigns of the 14 *manus* together span $14 \cdot 71 = 994$ *mahā-*

yugas and the 15 junctures span $15 \cdot \frac{4}{10} = 6$ *mahāyugas*. The sum of these durations gives us the total duration of Brahmā's day, the *kalpa*.

As we shall see later in verses 18–24, the planets make a whole number of revolutions (a revolution of a planet is its journey around the earth, from a given point with respect to the fixed stars to the same point) during a *mahāyuga*, but this is not the case with the apogees and nodes. However, during a *kalpa*, the planets, the apogees, and the nodes all make a whole number of revolutions. In other words, the *kalpa* corresponds to the Platonic idea of the Great Year, a period at the end of which everything returns to the same positions they had at its beginning.

(4–5) **In this day of Brahmā, 6 *manus*, 27 [*mahā*]*yugas*, three parts of a [*mahā*]*yuga*, and 3,179 [*saura*] years of the *kali*[*yuga*] [that is, the fourth part] had passed at the commencement of the *śaka* [era] of the good Śālivāhana.**

[When] the [47,400 divine] years mentioned earlier multiplied by 360 [are] subtracted from what has elapsed of the day of Brahmā, [we get the years] that have elapsed since [the commencement of] planetary motion. They are 1,955,883,179 at the commencement of the *śaka* [era]. [These 1,955,883,179 years] along with the [elapsed years] of the *śaka* [era] are the given years [that is, the years from the commencement of planetary motion to the present].

Jñānarāja now proceeds to tell us how much of the present *kalpa* has elapsed. The *śaka* era, which is said to have been instituted by King Śālivāhana, began in 78 CE according to our calendar. At the commencement of the *śaka* era, Jñānarāja tells us, the reigns of 6 *manus*, as well as 27 *mahāyugas*, a *kṛtayuga*, a *tretāyuga*, a *dvāparayuga*, and 3,179 *saura* years of a *kaliyuga* of the seventh *manu* have elapsed. Note that we are currently in a *kaliyuga*.

The 6 *manus*, including their seven corresponding junctures, spanned a period of

$$6 \cdot 71 \cdot 4{,}320{,}000 + 7 \cdot 1{,}728{,}000 = 1{,}852{,}416{,}000 \qquad (12)$$

saura years.

The 27 *mahāyugas*, the *kṛtayuga*, the *tretāyuga*, and the *dvāparayuga* that have elapsed during the reign of the seventh *manu* until the beginning of our present *kaliyuga* spanned a period of

$$27 \cdot 4{,}320{,}000 + \frac{9}{10} \cdot 4{,}320{,}000 = 120{,}528{,}000 \qquad (13)$$

saura years.

Finally, 3,179 *saura* years have elapsed from the beginning of the *kaliyuga* to the commencement of the *śaka* era, giving a total of

$$1{,}852{,}416{,}000 + 120{,}528{,}000 + 3{,}179 = 1{,}972{,}947{,}179 \qquad (14)$$

saura years from the beginning of the *kalpa* to the commencement of the *śaka* era.

However, according to the *saurapakṣa*, the school of astronomy that Jñānarāja follows, planetary motion did not commence at the begin of the *kalpa*. Rather, as Jñānarāja explained in *Siddhāntasundara* 1.1.20, a period of 47,400 divine years, called *śṛṣṭikāla* (literally, "time of creation"), passed from the beginning of the *kalpa* until the commencement of planetary motion. A divine year is a year of the gods. According to the Indian tradition, such a year equals 360 *saura* years, so the 47,400 divine years equal

$$360 \cdot 47{,}400 = 17{,}064{,}000 \qquad (15)$$

saura years. When this period is subtracted from the above result, we get the *saura* years that have elapsed between the commencement of planetary motion and the commencement of the *śaka* era, namely,

$$1{,}972{,}947{,}179 - 17{,}064{,}000 = 1{,}955{,}883{,}179 \qquad (16)$$

saura years, as stated in the verse.

After the 17,064,000 *saura* years of the *śṛṣṭikāla* have elapsed, all the planets, apogees, and nodes commence their motion starting from Aries 0°, the beginning of the sign Aries (see verse 6).

Why does the *saurapakṣa* claim that a period of 17,064,000 *saura* years has elapsed between the beginning of the *kalpa* and the commencement of planetary motion? There is no reason for this to be found in the religious and mythological texts of India. In fact, the idea is a mathematical trick. By insisting that such a period elapsed before planetary motion commenced, one is ensured that a mean planetary conjunction occurs at the beginning of our present *kaliyuga*.[173] Note that other numbers than 17,064,000 can be found that fulfill this requirement as well, and it is not clear why this particular number was chosen.

Note that the *śṛṣṭikāla* prevents the *kalpa* from acting fully as a Great Year. At the end of the *kalpa*, when there is a partial destruction of the universe, the planets, the apogees, and the nodes will only have moved for 4,320,000,000 – 17,064,000 *saura* years, and will thus not be back at their original positions at Aries 0°.

(6) **[When the longitudes of] the planets, apogees, and nodes computed for the beginning of a [given] *saura* year [are] diminished by the motion [of the given planet, apogee, or node] in a year multiplied by the *saura* years elapsed [since the commencement of planetary motion], they are situated at the beginning of Aries.**

What the verse is essentially saying is that when planetary motion began at the end of the *śṛṣṭikāla*, the position of each planet, each apogee, and each node was Aries 0°.

Let m be a positive integer. In the following we will use the mathematical notation that $a \equiv b \bmod m$ means that a and b differ only by a multiple of m, that is, that $a - b = k \cdot n$ for some integer k. For example, $737 \equiv 17 \bmod 360$ since $737 - 17 = 2 \cdot 360$.

Suppose that exactly n *saura* years have elapsed since the commencement of planetary motion, and let v be the mean motion of a given planet during a *saura* year. During the n *saura* years of motion, the mean planet moved the angular distance $n \cdot v$ (measured in degrees), which thus, except for a multiple of 360°, equals $\bar{\lambda}$, the planet's mean longitude (that is, the longitude of the mean planet) at that time. In other words, there is an integer k such that

$$n \cdot v = k \cdot 360° + \bar{\lambda}, \tag{17}$$

[173]See Pingree 1978a, 609.

or, equivalently,

$$\bar{\lambda} \equiv n \cdot v \bmod 360. \tag{18}$$

This tells us that $\bar{\lambda} - n \cdot v$ gives us the starting point of motion, that is, the zero point on the ecliptic. What the verse further tells us is that the starting point is the same for all of the planets, the apogees, and the nodes, namely, Aries 0°.

(7) **It has been said that the planets, being simultaneously in one place, commence their eastward motion at the beginning of Brahmā's day. Those [who say this] are opposed to the *vedas*, [because the idea] differs from the opinion of Brahmā, Sūrya, Candra, and others.**

The idea of the *śṛṣṭikāla*, used in the *saurapakṣa* to ensure a mean conjunction of the planets at the beginning of our *kaliyuga*, is not universally agreed upon in the Indian astronomical tradition. The *brāhmapakṣa*, the oldest of the schools of classical Indian astronomy, does not employ it, but rather has the planets, the apogees, and the nodes commence their motion right at the beginning of the *kalpa*. Here Jñānarāja attacks this opinion for being opposed to the opinion of divine personages recorded in the sacred texts.

The authorities whose opinions Jñānarāja appeals to here are the gods considered to be the narrators of the astronomical texts considered divinely and scientifically authoritative by Jñānarāja: the *Brahmasiddhānta*, the *Sūryasiddhānta*, the *Somasiddhānta*, and so on. These texts all belong to the *saurapakṣa*, and advocate the idea of the *śṛṣṭikāla*.

(8) **The pure [teaching] that Brahmā spoke to Nārada, Candra spoke to Śaunaka, the sage Vasiṣṭha spoke to Māṇḍavya, and Sūrya spoke to Maya is full of reasoning based on perception and traditional teachings. Whatever men do differently after abandoning this science, that ceases to produce correct results over time, because they are devoid of [proper] knowledge.**

An astronomical model with errors in the parameters or elsewhere might prove fairly accurate around the time when it was established, but over long periods of time the errors will reveal themselves as predicted planetary positions and so on start to deviate more and more from what is observed. Jñānarāja maintains that such will be the case when men deviate from the science as taught by gods and sages, their teaching being not only based on the sacred tradition but also full of sound reasoning.

The knowledge given by Brahmā to Nārada, by Candra to Śaunaka, by Vasiṣṭha to Māṇḍavya, and by Sūrya to Maya constitutes the contents of the treatises known as the *Brahmasiddhānta*, the *Somasiddhānta*, the *Vasiṣṭhasiddhānta*, and the *Sūryasiddhānta*, respectively.[174]

(9) **If somewhere something different from what is taught by the sages is perceived by men, then that alone is to be corrected. Everything is not to be done differently.**

Jñānarāja now opens up the possibility that some defect might be found in the teachings of the ancient sages. If that should happen, one is not to reject the teachings entirely, though, but rather correct the defect while maintaining the overall system and theory.

(10) **Just as [when] no strength is found somewhere in the *mantra*s enunciated in the *veda*s, one should perform their *puraścaraṇa*. Everything is not to be done differently.**

A *puraścaraṇa* is a procedure for making a *mantra* (a sacred verse, phrase, or mystical syllable, usually in Sanskrit, used for spiritual practice) effective.[175] The main part is repetition of a *mantra*, but it also includes other things, such as fire offerings and feeding of *brāhmaṇa*s (the class of priests and intellectuals in the Hindu tradition). It is a practice from the tradition of the *tantra*s (not to be confused with the astronomical treatises known as *tantra*s, the *tantra*s mentioned here are esoteric texts outlining various spiritual practices; see also commentary

[174]See Introduction, page 26.

[175]I am indebted to Gudrun Bühnemann, Christopher Minkowski, and Frederick Smith for explaining the idea of the *puraścaraṇa* via personal communications.

on *Siddhāntasundara* 1.1.2) known in both Hindu and Buddhist contexts, but the *mantra*s employed in the procedure can also be *mantra*s from the *veda*s.[176] Around the time of Jñānarāja, the practice of *puraścaraṇa* gravitated into the ritual world of the *veda*s.

Jñānarāja's point here is that when a *mantra* from the *veda*s is seen to lack strength or energy (*vīrya*), that is, if it is seen to be ineffective for some reason, it is not rejected. Rather, a *puraścaraṇa* of the *mantra* is carried out in order to make it effective. A *puraścaraṇa* can, for example, be carried out to infuse *mantra*s from the *veda*s with power via methods from the *tantra*s. In this example, *mantra*s from the *veda*s are considered not to be independently powerful, but require charging through Tantric practices. So, similarly, if a defect is found somewhere in the science of astronomy, one ought not to discard the whole science, but rather try to correct the defect.

(11) **The knowledge which is the opinion of the sages, which since ancient times has constantly been perceived as agreeing with observation, and which is to be understood through *vāsanā*s, that is agreed to by us.**

This verse paraphrases Jñānarāja's approach to astronomy. The science that he subscribes to is one that comes through the authority of ancient sages, but beyond being merely the opinion of authoritative figures, Jñānarāja holds that it agrees with observation and can be understood through various methods. As such, this knowledge is not mystical.

(12–17) **In a [*mahā*]*yuga* there are** 1,593,336 **intercalary months and** 25,082,252 **omitted *tithi*s. The number of lunar months [during a given period of time] is the difference between the revolutions of the sun and the moon [in that same period]. During a [*mahā*]*yuga*, [the number of lunar months] is given by** 53,433,336 **according to the sages. There are** 1,555,200,000 ***saura* days in a [*mahā*]*yuga*. The number of *tithi*s [in a *mahāyuga*] is** 1,603,000,080. **In a [*mahā*]*yuga* there are** 1,577,917,828 **civil**

[176] For the *puraścaraṇa* in the context of the *tantra*s, see Bühnemann 1991, 301–305 and Bühnemann 1992.

days and 1,582,237,828 sidereal days.

The knowledgable ones always say that the omitted *tithis* [in a *mahāyuga*] are the difference between the civil days and the *tithis*, and that the *tithis* in the intercalary months are the difference between the number of *saura* days and the number of *tithis*. The number of days of a planet [that is, the time between two consecutive risings of the planet] [in a *mahāyuga*] is measured by the revolutions of the stars diminished by the revolutions [of the planet in a *mahāyuga*]. In this case, because it is widely known, the method is not given; it is [arrived at] by intellect and one's own understanding alone.

In our modern calendar we insert a leap year every four years in order to keep the calendar year and the astronomical year synchronized. This is necessary because astronomical events do not repeat after an integer number of days, and hence a drift occurs between the event and the calendar day on which it occurs. For example, the season of spring would commence at different dates over time. The Indian tradition inserts an intercalary month at times for the same purpose, though this procedure is not done after as simple a scheme as in our calendar.

A lunar month is the time between one conjunction of the sun and the moon and the next. There are 30 *tithis* in a lunar month, the first ending when the moon has gained 12° over the sun in longitude, the second when the moon has gained a further 12°, and so on. Since the velocities of the sun and the moon vary, the duration of a *tithi* is not constant. The duration of a *tithi* varies between about 22 hours and about 26 hours.

There are 12 *saura* months of the same duration in a *saura* year, and 30 *saura* days, again of the same duration, in a *saura* month.

In the Indian astronomical tradition, a civil day is either time between two consecutive sunrises or between two consecutive midnights. Jñānarāja follows a midnight system, and hence the latter is the case in the *Siddhāntasundara*. A civil day is roughly 24 hours.

When a *tithi* occurs entirely between one midnight and the next (or between one sunrise and the next, depending on which astronomical school one follows), it is called omitted.

Unit	Symbol	Number in a *kalpa*
Intercalary months	A	1,593,336
Omitted *tithis*	U	25,082,252
Lunar months	M	53,433,336
Saura days	S	1,555,200,000
Tithis	T	1,603,000,080
Civil days	C	1,577,917,828
Sidereal days		1,582,237,828

Table 2: Astronomical units

The units in a *mahāyuga* presented here by Jñānarāja are given in Table 2, together with the symbols that we will use to designate them in the following.[177] The units are based on the number of revolutions of the planets which he will give subsequently in verses 18–24.

Let us assume that two planets, P_1 and P_2, make r_1 and r_2 revolutions, respectively, during a given period of time, and that P_1 is the faster of the two. If they are at the same position at the beginning of the period, the faster planet will catch up with the other planet $r_1 - r_2$ times during the period.

We can prove this result mathematically as follows. Suppose that the planet P_1 moves with the velocity $v + u$ (measured in degrees per unit time), making r_1 revolutions in a period of time of length t. Suppose similarly that the planet P_2 moves with the velocity v, making r_2 revolutions in the same period of time. Finally, suppose that the two planets are at the same position at the beginning of the period. Then $t \cdot (v + u) = 360 \cdot r_1$ and $t \cdot v = 360 \cdot r_2$, and therefore $t \cdot u = 360 \cdot (r_1 - r_2)$. In order for the two planets to have the same position at a time τ with $0 \leq \tau \leq t$, we must have $\tau \cdot (v + u) \equiv \tau \cdot v \bmod 360$, that is, that there is a positive integer k, so that $\tau \cdot u = 360 \cdot k$. This means that if n is chosen so that $360 \cdot n \leq t \cdot u < 360 \cdot (n+1)$, P_1 catches up with P_2 a total of n times during the given period of time. But $t \cdot u = 360 \cdot (r_1 - r_2)$, so $n \leq r_1 - r_2 < n + 1$, and n is therefore the integer part of $r_1 - r_2$. This completes the proof.

As a lunar month is the period from one conjunction of the sun and the moon to the next, we can use this result to find the

[177] The symbols are the same as those used in Pingree 1978a.

number of lunar months in a *mahāyuga*. According to verse 18 below, the sun and the moon make, respectively, 4,320,000 and 57,753,336 revolutions during a *mahāyuga*. Hence, there are

$$57{,}753{,}336 - 4{,}320{,}000 = 53{,}433{,}336 \tag{19}$$

lunar months in a *mahāyuga*, as stated in the verse.

Since a *saura* year is one sidereal revolution of the sun starting at Aries 0° and a *mahāyuga* consists of 4,320,000 revolutions of the sun, there are 4,320,000 *saura* years in a *mahāyuga*. A *saura* year comprises 12 *saura* months, each of which, in turn, comprises 30 *saura* days. Hence, there are

$$30 \cdot 12 \cdot 4{,}320{,}000 = 1{,}555{,}200{,}000 \tag{20}$$

saura days in a *mahāyuga*.

Since there are 53,433,336 lunar months in a *mahāyuga*, there are $30 \cdot 53{,}433{,}336 = 1{,}603{,}000{,}080$ *tithi*s in a *mahāyuga*.

The number of civil days is found as the revolutions of the stars in a *mahāyuga* diminished by the revolutions of the sun in a *mahāyuga*. The revolutions of the stars in a *mahāyuga* are, of course, the number of sidereal days in a *mahāyuga*.

The omitted *tithi*s are the excess of *tithi*s compared to civil days in a *mahāyuga*, which is $1{,}603{,}000{,}080 - 1{,}577{,}917{,}828 = 25{,}082{,}252$.

We can find the number of *tithi*s in the intercalary months in a *mahāyuga* as the difference between the *tithi*s in a *mahāyuga* and the *saura* days in a *mahāyuga*. These are $1{,}603{,}000{,}080 - 1{,}555{,}200{,}000 = 47{,}800{,}080$. The number of intercalary months can then be found as $\frac{47{,}800{,}080}{30} = 1{,}593{,}336$.

Civil days are determined by the sun's progress around the earth. We can define days of other planets in precisely the same way. If we do so for a given planet, the number of days of that planet in a *mahāyuga* are found exactly like the number of civil days in a *mahāyuga*; we just use the revolutions of that planet rather than the revolutions of the sun.

(18–24) **In a [*mahā*]*yuga* there are** 4,320,000 **revolutions of the sun, Mercury, and Venus. The revolutions of the moon are given by the wise as** 57,753,336. **The number of revolutions of Mars in a [*mahā*]*yuga* is** 2,296,832. **The revolutions of Mercury's *śīghra* are** 17,937,060. **The**

number of revolutions of Jupiter is considered to be 364,220. The number of revolutions of Venus's *śīghra* in a [*mahā*]*yuga* is 7,022,376. The number of revolutions of Saturn is 146,568. The number of revolutions of the lunar apogee is 488,203. [All of the these move] with an eastward motion in the sky. In the case of the moon's node, [known as] Rāhu, the revolutions are understood to be 232,238 in the opposite direction. These [numbers] multiplied by 1,000 give [the revolutions] during a *kalpa*.

Now, the revolutions of the sun's *manda* apogee in a *kalpa* are 387. Beginning with Mars, [the revolutions] produced by the *manda* apogees [of the star-planets] are 204, 368, 900, 535, and 39.

In a *kalpa* [the revolutions] of the nodes, which are moving in the opposite direction, are, beginning with Mars, 214, 488, 174, 903, and 662.

These verses give the number of revolutions of each planet, each apogee, and each node during either a *mahāyuga* or a *kalpa*.

When authors of Indian astronomical texts use the phrase "starting from Mars," the order is understood to be Mars, Mercury, Jupiter, Venus, and Saturn. This is the order of the days of the week, Tuesday being the day of Mars, Wednesday the day of Mercury, Thursday the day of Jupiter, Friday the day of Venus, and Saturday the day of Saturn. Sunday is the day of the sun, and Monday the day of the moon.

Now, the number of revolutions given here signifies the revolutions of the mean planets. The mean planet is not something found in reality, but is a construct used in the astronomical model. Each planet moves with variable speed around the earth, but each planet also has an average velocity. Assume that there is a body moving with this constant velocity around the earth along a perfect circle and that the body and the actual planet were at the same position when planetary motion began. Then this body is the mean planet corresponding to the planet in question.

The planetary theory of Indian astronomy will be explained by Jñānarāja in the next section. For now it suffices to say that each of the star-planets has two epicycles, each with an apogee. These two epicycles are called *manda* (slow) and *śīghra* (fast).

Body	Revolutions per *mahāyuga*	Revolutions per *kalpa*
The sun	4,320,000	4,320,000,000
manda apogee		387
The moon	57,753,336	57,753,336,000
manda apogee	488,203	488,203,000
node	−232,238	
Mars	2,296,832	2,296,832,000
manda apogee		204
node		−214
Mercury	4,320,000	4,320,000,000
śīghra apogee	17,937,060	17,937,060,000
manda apogee		368
node		−488
Jupiter	364,220	364,220,000
manda apogee		900
node		−174
Venus	4,320,000	4,320,000,000
śīghra apogee	7,022,376	7,022,376,000
manda apogee		535
node		−903
Saturn	146,568	146,568,000
manda apogee		39
node		−662

Table 3: Revolutions of the planets, apogees, and nodes

We can think of the *manda* epicycle as accounting for the fact that the planets do not move around the earth in perfect circles, and the *śīghra* epicycle as accounting for the fact that the earth orbits the sun rather than the sun orbiting the earth.

Table 3 shows the revolutions as given by Jñānarāja. They are identical with those given in the *Sūryasiddhānta*.[178]

(25) **These are the very revolutions agreed upon in the *tantras* [treatises] [composed] by Brahmā, Sūrya, Candra, Vasiṣṭha, Pulastya, and so on.**

[178] See Pingree 1978a, 608, Table VIII.1; 609, Table VIII.5 for tables of the revolutions according to the *Sūryasiddhānta*.

Here [in this work] I will now present a method that is pure, easy, unprecedented, and that provides understanding of the computation of the revolutions.

Dikshit observes that the revolutions that are given in the *Sūryasiddhānta*, the *Somasiddhānta*, the *Vasiṣṭhasiddhānta*, the *Romakasiddhānta*, and the *Brahmasiddhānta* (the astronomical treatises spoken by Sūrya, Candra, Vasiṣṭha, Viṣṇu, and Brahmā, respectively) are identical and lists them in a table.[179] With the exception of the revolutions of the node of Saturn, which Dikshit gives as 60, the values agree with those given by Jñānarāja. However, the 60 revolutions given by Dikshit for the node of Saturn may be a misprint, as the *Sūryasiddhānta* gives 662 revolutions like Jñānarāja.[180]

As has been noted, Jñānarāja is concerned with providing demonstrations that establish the validity of the divine knowledge that he is presenting. In keeping with that concern, a method by which the revolutions can be found through observation is given in the following.

(26) **The shadow of a gnomon that is straight and positioned on ground that has been made even by means of water [falls along the] south-north [line] at [the time of] the *ghaṭikās* of midday. The east and west directions are produced from the tail and head of [a figure in the shape of] a fish produced from it [the north-south line].**

The straightforward construction described determines the cardinal directions. It is based on the fact that at noon, that is, when the sun is on the local meridian (that is, the great circle connecting the North and the South Poles and passing through the local zenith), the shadow cast by a gnomon will be aligned along the north-south line.

In Figure 5, G is the gnomon and A is the tip of the shadow cast by the gnomon at noon. The line GA is then the north-south line. Drawing two intersecting circles with the same radius centered around G and A, respectively, let B and C be the two points of intersection between the circles. The line BC is then the

[179] See Dikshit 1969–81, 2.27–28.

[180] *Sūryasiddhānta* 1.44. See also Pingree 1978a, 608, 609, Table VIII.5.

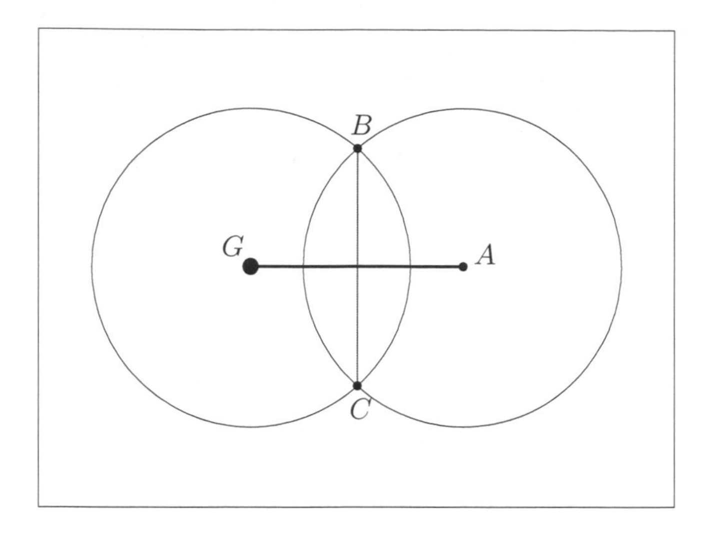

Figure 5: The east-west and north-south lines

east-west line. We have now determined the cardinal directions. The intersection of the two circles, which has the appearance of a fish, is called a "fish figure" in Indian astronomy. So, here a fish figure is used to determine the east-west line.

Jñānarāja does not directly address the question of how to determine when precisely it is noon, but by expressing "noon" as "the *ghaṭikā*s of midday," he hints that this is to be done by keeping track of time. If we know how many *ghaṭikā*s there are between noon and (presumably) sunrise on the day in question, we can use a water clock to determine when it is noon.

This verse is given again later in the text by Jñānarāja (*Siddhāntasundara* 2.3.2).

(27–29[a]) **On the south-north [line] one should place the bottom edge of a post that is made from beautiful wood, [so that the post] is perfectly perpendicular and very straight.**

[First,] having fixed a peg on the upper part [of the post] and a flexible sighting tube on the peg [so that it falls along the] south-north [line], then, observing the sun with that [sighting tube], determining the number of degrees of the altitude [of the sun] from an instrument such as the *turya-yantra* [that is, the sine quadrant], and

subtracting [this number] from 90 [degrees], the degrees of the zenith distance [of the sun] is the result. The difference or sum of that [zenith distance] and one's own latitude, when the directions are the same and different, respectively, is the degrees of the declination [of the sun].

The setting up of the sighting tube is clear from the verse. When the sighting tube is aligned with the north-south line, it can be used to observe the sun when it is on the meridian. However, looking directly at the sun through the sighting tube could damage the eyes of the observer, so perhaps the intended procedure is to let sunshine pass through the sighting tube onto the surface of water in a vessel.

A sine quadrant (*turya-yantra*) is an astronomical instrument used to measure the altitude of the sun, that is, the angular distance of the sun above the horizon.[181] The instrument is discussed by Jñānarāja in *Siddhāntasundara* 1.5.3 and the following verses. Presumably, when the sighting tube points to the sun, its orientation is used to determine the altitude of the sun via the quadrant.

Let α be the altitude of the sun. The sun's "zenith distance" z is its angular distance from the zenith, which is 90° above the horizon, so $z = 90^\circ - \alpha$. The local latitude ϕ equals the angular distance of the zenith above the celestial equator, while the sun's declination δ represents its own angular distance from the equator. Thus, δ is given by combining z and ϕ, as described below.

This is illustrated in Figure 6. In both the analemmas of the figure the circle is the meridian, the horizontal line the local horizon, the vertical line the prime vertical, and the oblique line through the entire circle the equator. In both cases, the sun, being the point S, is on the meridian. The point Z is the local zenith, and the point E is the intersection between the celestial equator and the local meridian. The point O is the center of the earth.

In the first case, the sun's zenith distance is the angle SOZ, and the local latitude is the angle ZOE. It is clear that the sun's

[181] See Evans 1998, 205–206.

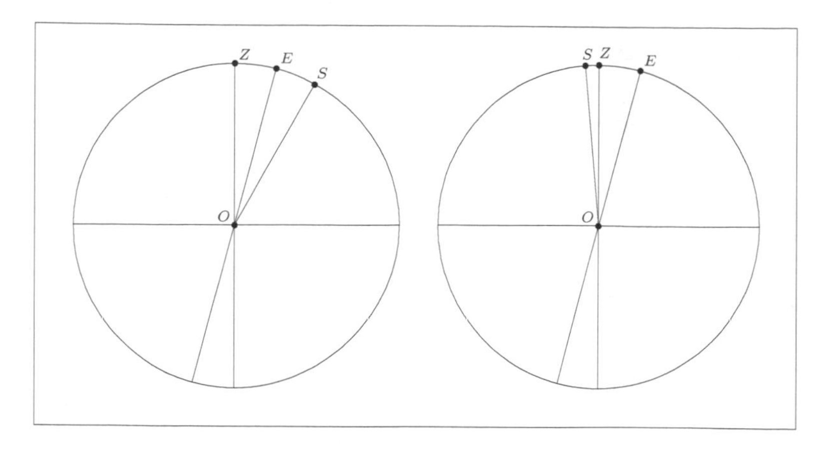

Figure 6: The sun's declination from its zenith distance

declination, δ, is

$$\delta = \angle SOE = \angle SOZ - \angle ZOE = z - \phi, \tag{21}$$

that is, the sun's zenith distance diminished by the local latitude. In the second case, we similarly have that

$$\delta = \angle SOE = \angle SOZ + \angle ZOE = z + \phi. \tag{22}$$

In this way the declination of the sun can be found from its zenith distance and the local latitude.

($29^{\text{b–d}}$) **[When] the Sine of that [declination] is divided by the Sine of 24 [degrees] and multiplied by the Sine of 90 [degrees], the arc [corresponding] to that [resulting quantity] is the degrees of the arc of the sun [measured from the vernal equinox, that is, the sun's tropical longitude] on that day.**

At the beginning of the night, after observing a star, one should fix its [the sun's] position.

The first part of the verse gives a formula for computing the tropical longitude of the sun (that is, the longitude of the sun measured with respect to the vernal equinox, which is the intersection between the ecliptic and the celestial equator at which the sun crosses from the southern to the Northern Hemisphere) from its declination and the obliquity of the ecliptic. The formula involves the sine function, which will be properly introduced by Jñānarāja in the beginning of the next section. For now it suffices to say that the Indian tradition operates with a sine function with a non-unity trigonometrical radius. If α is any angle, we will denote the Indian sine of α by $\mathrm{Sin}(\alpha)$. This equals $R \cdot \sin(\alpha)$, where R is the trigonometrical radius (Jñānarāja uses $R = 3{,}438$; for more information, see the commentary on *Siddhāntasundara* 2.2.2–5). When speaking of an Indian sine, we will write "Sine" rather than "sine."

The obliquity of the ecliptic is the angle between the ecliptic and the celestial equator. It is taken to be 24° in the Indian tradition and is what the 24° of the verse refers to. It is denoted by ε in the following.

If λ^* is the tropical longitude of the sun and δ the declination of the sun, the formula given can be written as

$$\mathrm{Sin}(\lambda^*) = \frac{R \cdot \mathrm{Sin}(\delta)}{\mathrm{Sin}(\varepsilon)}. \tag{23}$$

This is, of course, simply the formula known as the "method of declinations," that is, $\mathrm{Sin}(\delta) = \frac{\mathrm{Sin}(\lambda^*) \cdot \mathrm{Sin}(\varepsilon)}{R}$, by which the declination of the sun can be found from its tropical longitude and the obliquity of the ecliptic (that is, the angle between the ecliptic and the celestial equator).

The idea of the last part of the verse seems to be as follows. At sunset, when the sun is setting on the western horizon, one should observe what star is rising in the east. The sun will be 180° from that star, so the sun's sidereal longitude can be determined in this way.

(30) **In this way, having first observed the precessional [longitude of the] true sun [on a given day], next [the same procedure is to be followed] on another day. The true motion [of the sun between the two days] is the difference between the two [longitudes of] the sun. As much as the [longitude of] the sun is when that [true motion] is at a minumum, [the longitude of] its apogee is equal to that.**

By precessional (*sāyana*) longitude is meant the sun's longitude corrected for precession, that is, its tropical longitude.

If we observe the sun at noon on two consecutive days, we get two tropical longitudes. The difference between these is the angular distance traveled by the sun between noon on the first day and noon on the second day. Continuing this process, a table of daily velocities of the sun is produced. On the day that the sun's velocity is at its minimum, it is understood that the sun is at its apogee, that is, it is the farthest from the earth that it gets.[182] In other words, the longitude of the solar apogee can be found as the longitude of the sun when its minimum daily velocity is attained.

(31–32) **The mean motion [of the sun] is half of the sum of the smallest and the greatest true motions.**

When the greatest motion occurs, then [the longitude of] the mean sun is equal to the true [longitude].

Having determined that [longitude of the mean sun] in signs and so on, then each day one should make the mean [sun] move with the mean velocity according to the interval of time [elapsed from the time corresponding to the known mean longitude].

The equation [of the sun] is the difference between its mean and true [longitudes].

The mean sun is a theoretical construct forming part of the model of solar motion. It moves with constant velocity on a circle with the earth as its center.

Let v_1 and v_2 be the maximum and minimum velocities of the sun, respectively, and let $\overline{v}$ be the mean velocity of the sun. Jñānarāja takes the mean velocity to be the average of the maximum velocity and the minimum velocity:

$$\overline{v} = \frac{v_1 + v_2}{2}. \tag{24}$$

We saw before that when the sun travels with its minimum velocity, it is at its apogee. Similarly, when it travels with its maximum velocity, it is at its perigee (the point opposite, that is, 180° from, the apogee). At both the apogee and the perigee

[182]For the model governing solar motion and the reason for the solar velocity being at its minimum when the sun is at its apogee, see the next section on true motion.

(Jñānarāja only specifies the perigee indirectly by saying that the solar velocity is maximal), the sun's mean longitude is equal to its true longitude. As we know the true longitude, we now have a point in time where the sun's mean longitude is known. Let this longitude be $\bar{\lambda}$.

Starting with this point in time, we can now compute the position of the mean sun at any given time. The mean sun moves with the constant velocity $\overline{v}$, so if a period of time of length t has elapsed since the point in time at which we knew the sun's mean longitude, the mean longitude after this period of time has elapsed is $\bar{\lambda} + t \cdot \overline{v}$.

Knowing both the true and the mean longitude of the sun at a given time, we can find the equation, which is the difference between the two (this will be discussed in greater detail in the next section).

(33) **Wherever there is equality of the mean and true velocities, [there] is the greatest equation.**

It is to be understood that the radius of the epicycle is equal to the Sine of that [greatest equation].

The greatest equation is the greatest possible angular distance between the true planet and the mean planet. It occurs when the sun is traveling with its mean velocity and is roughly equal to the radius of the solar epicycle.

(34) **In this way, as great as the [longitude of] the sun is in signs, degrees, and so on, so great is it owing to the [passing of civil] days and their parts [such as *ghaṭikās*], [these elapsed days being] referred to as "star-eaten."**

If one revolution is achieved by means of [a certain number of] these [civil days], then [how many revolutions] are achieved by means of the civil days in a *kalpa*? [It is explained] by the wise that those [revolutions in the result] are the revolutions [of the sun] in a *kalpa* computed thus from a proportion.

The position of the sun changes with time, elapsed time being called "star-eaten," though the exact interpretation of this is unsure.

We can find the time it takes for the sun to make one revolution. Say that this takes a civil days. Let C be the number of civil days in a *kalpa*. Then

$$\frac{1}{a} = \frac{r}{C}, \tag{25}$$

where r is the number of revolutions of the sun in a *kalpa*. In this way, the number of solar revolutions in a *kalpa* can be determined by means of a proportion.

(35–36[a–b]) **Having observed the moon when it is on the meridian with a quadrant as explained earlier, it is to be understood that its declination is corrected for the lunar latitude.**

Regarding the corrected *ghaṭikā*s to be the lapsed *ghaṭikā*s of the night at that time [found] by means of an instrument, the accurate [longitude of the] meridian ecliptic point is to be computed from [the longitude of] the sun by means of the rising times at Laṅkā using the method given subsequently.

It is said by the ancients that the moon with the visibility correction applied is equal to that [longitude of the meridian ecliptic point].

The declination of the moon is the angular distance between the celestial equator and the moon's position on the ecliptic. However, the moon is not on the ecliptic, but rather on its own inclined orbit. The angular distance between the ecliptic and the moon's position on its inclined orbit is called the lunar latitude. When the moon is observed in the sky, what is seen is its actual position on its inclined orbit. Hence, as is stated, the moon's declination is corrected for the lunar latitude; its angular distance to the celestial equator is the combination of its declination and the lunar latitude.

If we know how many *ghaṭikā*s have elapsed of the night when the moon is seen on the meridian, for example by measuring the time with a water clock, we can compute the longitude of the sun at that time, and from that the longitude of the meridian ecliptic point (the intersection of the meridian and the ecliptic) can be found by means of the rising times of the signs (given by Jñānarāja in *Siddhāntasundara* 2.2.39–40).

When the moon is seen on the meridian, its position with corrections applied is that of the longitude of the meridian ecliptic point.

(36^{c-d}) **The latitude of the moon [when on the meridian] is the difference between the corrected declination [of the moon] and the declination of the meridian ecliptic point.**

The declination of a point is its angular distance to the ecliptic. The meridian ecliptic point is, of course, on the ecliptic, but the moon is not. The moon is on its own inclined orbit. The angular distance between the moon and the ecliptic is called the lunar latitude. Jñānarāja notes that in the given situation, the lunar latitude can be found as the difference between the declinations of the meridian ecliptic point and the corrected declination of the moon (corrected in the sense that lunar latitude and so on have been applied). This is correct if the meridian ecliptic point is considered to coincide with the nonagesimal, that is, the point on the ecliptic between the ascendant and the descendant (the ascendant is the point on the horizon where the ecliptic is rising at a given time, and the descendant is similarly the point where it sets at that time). However, as is most often the case, these two points do not coincide, in which case the declination of the meridian ecliptic point and the corrected declination of the moon have slightly different directions than the lunar latitude, and the formula is thus only approximately correct.

(37) **Whenever the southern latitude vanishes, [the longitude of] the moon subtracted from a rotation is [the longitude of] the [ascending] lunar node.**

Having again [at another time] determined [the longitude of] that node, its velocity can be computed from the difference in time [between the two observations].

What is meant here is that at the point in time when the southern latitude vanishes, the latitude becoming 0°, the moon crosses the ecliptic moving from the Southern Hemisphere to the Northern Hemisphere. There are two intersections between the inclined orbit of the moon and the ecliptic, both of which are called nodes. The node at which the moon crosses the ecliptic into the Northern Hemisphere is called the ascending node, the other the descending node. If we know the longitude of the moon when it is at the ascending node, we can find the longitude of the ascending node by subtracting the longitude of the moon from a rotation, that is, 360°. The reason for the subtraction is that the node moves in the opposite direction of the moon.

If this procedure is carried out twice, we can find the velocity of the ascending node.

(38) **The *manda* apogee of the moon is to be found like the apogee of the sun, and likewise its [the moon's] [mean] motion. The revolutions [of the moon] in a *mahā-yuga* or a *kalpa* are to be computed by means of this [motion].**

Following the same procedure as for the sun, the longitude of the *manda* apogee of the moon can be found, and from that the mean velocity of the moon. The number of revolutions of the moon in a *mahāyuga* or in a *kalpa* can then be found from that mean velocity.

(39) **Going ahead as before, the number of days in one revolution of a [star-]planet from the beginning of its retrogradation [is to be found through observation]. [Then] one should compute half of the sum of the maximum and the minimum of that. [Finally,] one should compute [the star-planet's] revolutions in a *kalpa* by means of that [quantity].**

The normal direction of a planet is eastward. The sun and the moon always move eastward, but the five star-planets occasionally change their direction and go westward for some time. A planet that is traveling eastward is said to be in prograde motion. A planet that is traveling westward is said to be in retrograde motion. During prograde motion, the planet's longitude steadily increases, whereas it steadily decreases during retrograde motion.

When a planet is about to change its direction, it appears to stand still in the sky for a period of time. When an eastward-moving planet stands still in the sky before moving westward, it is said to be at its first station. Similarly, when a westward-moving planet stands still before moving eastward, it is said to be at its second station.

Through observation, one can find the number of days in a revolution of a star-planet from when it commenced its retrograde motion. Repeating this over time, one can find the mean velocity of the planet by taking the average of the largest and the smallest number of days found. In this way, the revolutions of the star-planet in a *kalpa* can be found.

(40–41) **The combination of [the effects of] the *manda* and *śīghra* equations is always the difference between the [star-]planet as established by observation and the mean [star-planet]. When [this combination] is corrected by the reverse of the *śīghra* equation, only the *manda* equation remains.**

Even though there is no [*śīghra*] equation when [the longitude of] a [star-]planet is equal to [the longitude of] its own *śīghra* perigee or apogee, the true [longitude of the star-]planet is not equal to the mean [longitude in this case]. Therefore, the *manda* apogee of the planet [contributes] a little bit else [to the planet's equation].

A star-planet has two epicycles. The difference between the mean planet and the observed planet is precisely the effect of these two epicycles. If the effect from the *śīghra* epicycle is removed, only the effect from the *manda* epicycle remains. Note, however, that removing the effect of the *śīghra* epicycle is generally not nearly as easy as Jñānarāja makes it out to be here.

The reason that the mean and true planets do not coincide even when there is no *śīghra* equation is that the *manda* epicycle also has an effect, albeit smaller than that of the *śīghra* epicycle.

(42) **There is no *śīghra* equation in the case of planets [whose longitudes] are equal to their own *śīghra* perigees or apogees. In that case, the true [planet] is the *manda*-corrected [planet], and the difference between that [*manda*-corrected planet] and the mean [planct] is the *manda* [equation].**

When the mean planet is corrected by the *manda* equation, it is called the *manda*-corrected planet. When there is no *śīghra* equation, the true planet is the *manda*-corrected planet, and the difference between the true and the mean planets is the *manda* equation.

(43) **When Saturn, Mars, and Jupiter are located in front of the sun, the longitude of the true planet is seen to be less than [the longitude of] the mean planet, [whereas] when they are located behind [the sun], it is seen as greater. Therefore, it is pointed out by the ancients that [the longitude of] the *śīghra* apogee of [these] three [planets] is equal to [the longitude of] the [mean] sun.**

In the case of the superior planets, that is, Mars, Jupiter, and Saturn, the longitude of the *śīghra* apogee is always that of the mean sun.[183]

($44^{\text{a–b}}$) **Knowing that when there is no *manda* equation, [the longitude of] a planet is equal to [the longitude of its] *manda* perigee or apogee, its measure is to be computed.**

It is not quite clear how the longitude of the planet is to be computed from the knowledge that it is equal to the longitude of the *manda* perigee or apogee when there is no *manda* equation.

($44^{\text{c–d}}$) **[The true longitudes of] Venus or Mercury are to be determined [when the planets are] on the horizon, [so that their longitudes are] equal to [the longitude of] the ascendant. Their velocities are to be computed from those [longitudes].**

When Mercury or Venus is on the horizon, their longitudes can be taken to be equal to that of the ascendant (as the two planets are on their own inclined orbits, their positions will generally not coincide with that of the ascendant, but we can as-

---

[183]See Pingree 1978a, 557.

sume that they approximately are). Repeating this observation, we can find their velocities.

(45) **For as long a time as [Mercury and Venus] are located in front of the sun, so long are they always located behind [it]. Therefore, for the sake of computing [the longitudes of] Mercury and Venus it is to be known that [their] revolutions during a *kalpa* are equal to the revolutions of the sun [in the same period of time].**

Since Mercury and Venus are seen as much in front of the sun as behind it, it can be inferred that their mean motion is identical to that of the sun. Hence, they have the same number of revolutions in a *kalpa* as the sun.

(46–49) **The number of *ghaṭikā*s elapsed since the rising of Venus is produced by means of its own shadow and by means of computation. Since that is what has elapsed of the day [of Venus] for a person at the center of the earth, so one's own [time] is easily computed by a water clock.**

If the *yojana*s measured by the radius of the earth [are attained] by means of the difference between the two elapsed [parts] of the day, then what [is attained] from the measure of *ghaṭikā*s in a nychthemeron? The result in this case is [the *yojana*s in] the orbit by means of a proportion.

Half of the sum of the maximum and minimum [values of] that [orbit] is the mean orbit of the *śīghra* apogee of Venus. The orbit of the sky divided by that is the number of revolutions of the *śīghra* apogee of Venus [in a *kalpa*].

In this very way one should [also] compute the revolutions of the *śīghra* [apogee] of Mercury. Thus is the excellent method taught. The revolutions of the *manda* [apogees], the *śīghra* [apogees], the [mean] planets, and the nodes are correct as per the words of the ancient sages.

The phrasing of verse 46 is not clear. However, given the

contents of verse 47, the meaning has to be the following. The time that Venus rises on the local horizon is not the same as the computed time, which is found with respect to the center of the earth. We have to find the time between the computed time and the time since rising on the local horizon.

The difference of these two times corresponds to Venus moving a distance roughly equal to the radius of the earth. This is completely analogous to the procedure described in *Siddhānta-sundara* 1.1.69–75. It is clear that the average between the greatest orbit and the smallest orbit yields the mean orbit. In addition, since we are dividing the orbit of the sky with the number found, we get the *śīghra* motion of the inferior planet in question.[184]

(50–51) **The number of years elapsed since the commencement of [planetary] motion is multiplied by 12 and increased by the elapsed months [of the current year]. [The result] is multiplied by 30 and increased by the [elapsed] *tithi*s [of the current month]. [This result is written down] separately [in two places]. [The first place] is multiplied by the given [number of] intercalary months [in a *mahāyuga*]. [The second place] is increased by the elapsed intercalary months multiplied by 30, [the elapsed intercalary months being found as] the result of the division of [the first result] by the *saura* days [in a *mahāyuga*]. [This result is written down] separately [in two places]. [The first place] is multiplied by the omitted *tithi*s in a [*mahā*]*yuga*. [The second place] is diminished by the elapsed omitted *tithi*s [found as] the result of the division of [the first result] by the [number of] *tithi*s [in a *mahāyuga*].**

This formula for computing the elapsed civil days between the commencement of planetary motion and the present is common and found in all *siddhānta*s.

Let y be the number of elapsed *saura* years between the commencement of planetary motion and the present, let n be the number of elapsed *saura* months between the commencement of planetary motion and the present, let m be the number

[184] See Pingree 1978a, 556.

of elapsed lunar months in the current year, let t' be the number of elapsed *tithi*s in the current month, let s be the number of elapsed *saura* days between the commencement of planetary motion and the present, let a be the number of elapsed intercalary months between the commencement of planetary motion and the present, let t be the number of elapsed *tithi*s between the commencement of planetary motion and the present, let u be the number of elapsed omitted *tithi*s between the commencement of planetary motion and the present, and let c be the number of elapsed civil days between the commencement of planetary motion and the present. We want to compute c.

The computation proceeds as follows:

$$12 \cdot y = n, \tag{26}$$
$$30 \cdot (n + m) + t' = s, \tag{27}$$
$$s \cdot \frac{A}{S} = a, \tag{28}$$
$$s + 30 \cdot a = t, \tag{29}$$
$$t \cdot \frac{U}{T} = u, \tag{30}$$
$$t - u = c. \tag{31}$$

Notice that in (27) we treat the elapsed lunar months as *saura* months and the elapsed *tithi*s of the current month as *saura* days, which is incorrect but always done in Indian astronomical texts. The computations in (29) and (31) follow from *Siddhāntasundara* 2.1.12–17.

(52) **The day count, which arises from the measure of the mean days of the sun, begins with a Sunday.**

One should subtract the part when it is larger from the residue of the omitted *tithi*s and the intercalary months [?].

A civil day is called a day of the sun because it is the time between two consecutive sunrises. The first day after the planets begin their motion is a Sunday.

The last half of the verse is not clear.

(53) **[When] the number of lapsed *saura* years [since the beginning of planetary motion] is multiplied by 12 and**

[then] increased by the [lapsed] months [of the current year] starting with *caitra*, [the result is the lapsed *saura* months since the beginning of planetary motion]. [When the number of lapsed *saura* months is] multiplied by 30 and [then] increased by the [lapsed] *tithi*s [of the current month], the [lapsed] *saura* days [since the beginning of planetary motion] are [the result]. Having computed the [lapsed] *adhimāsa*s [since the beginning of planetary motion] by means of those [lapsed *saura* days] put down in two places, [the lapsed *saura* days] are increased by those [lapsed *adhimāsa*s] converted into *tithi*s. [The result] is the [number of lapsed] *tithi*s [since the beginning of planetary motion], which is put down in two places. [The number of lapsed *tithi*s] diminished by the omitted *tithi*s [since the beginning of planetary motion], which are found by means of a proportion, is [the number of lapsed] civil [days since the beginning of planetary motion].

This is the same formula for computing the day count as was given in verses 50–51.

(54–55) **The beginning of the *saura* year [occurs] at [the sun's] entry into Aries. The epact is the difference between that [point in time] and the beginning of the month of *caitra*. By means of that [epact] subtracted from the beginning of the *saura* year, [we get,] in a different way, a day count commencing from the beginning of the bright [*pakṣa*] of *caitra*.**

The residue of the omitted *tithi*s is always found as the difference between midnight and the end of the *tithi*. By means of that [residue] subtracted from the time of [mid]night, [we get] a day count [commencing at the end of a given *tithi*], and the computation of [the mean longitudes of] the planets [can be accomplished] from that.

The month in which the sun enters the sign Aries is called *caitra*. The sun's entry into Aries marks the beginning of the *saura* year, and *caitra* is the first month of the year. In general, the beginning of *caitra* does not coincide with the sun's entry into the sign Aries. The time between the two is called the epact (*śuddhi*). When the epact is subtracted from the beginning of the *saura* year, that is, from the time of the sun's entry into Aries, we get the time of the beginning of *caitra*. This point in time can be used as the starting point for a day count instead of the commencement of planetary motion.

The remainder of the accumulated omitted *tithi*s is the difference between midnight and the end of the *tithi*. We can find the end of a given *tithi* from the time of midnight and knowledge of this remainder, and we can use this point in time for a day count. However, why one would want to use the end of a *tithi* for this purpose is unclear.

(56) **The result from [the division of] the given day count multiplied by the revolutions [of a given planet in a *mahāyuga*] by the civil days [in a *mahāyuga*] is the rotations and so on [of the planet during the day count].**

[The longitude of] the planet [found from a given day count] is [for] when it is midnight at [Laṅkā,] the city of the 10-headed [Rāvaṇa,] as determined by the mean sun.

Let r be the number of revolutions of the given planet in a *mahāyuga*. If d is the day count starting from the commencement of planetary motion, then it is clear that

$$360 \cdot r \cdot \frac{d}{C} \tag{32}$$

is the number of degrees traveled by the planet during the day count d. Subtracting the largest possible multiple of 360 from this, we get the planet's longitude.

The *saurapakṣa* employs a midnight system rather than a sunrise system. In other words, a day begins at midnight rather than at sunrise. Thus the longitude determined is the planet's longitude at midnight in Laṅkā on the day given by the day count.

(57) **The number of [elapsed] years in the *śaka* [era] of Śālivāhana is diminished by 1425. [The result] multiplied by the multiplier of a [given] planet and increased by the**

addend [of the planet] is the *dhruva* for the [present] year.

The multipliers and the addends are given in the following verses by Jñānarāja and are discussed in the notes on these verses. Put simply, the multiplier of a planet is the distance it travels during a *saura* year, and the addend of a planet is its position at a particular time in the year *śaka* 1425. The point in time corresponding to the addends of the planets is called the epoch.

Let g and a be the multiplier and the addend of a given planet. If n years have elapsed since *śaka* 1425, we compute $(n - 1425) \cdot g + a$. The result gives us the position of the planet exactly n *saura* years after the epoch. This position is called the planet's *dhruva* for that year.

As we shall see in the notes to the next verses, the epochal date corresponds to sunrise on a particular day. However, the positions found for an integral number of *saura* years after the epochal date do not correspond in a simple manner to sunrise (or any other given time of a nychthemeron) and are thus of limited use.

(58–64) **In the case of the moon, the multiplier in rotations and so on is** 4 **[signs],** 12 **[degrees],** 46 **[minutes],** 40 **[seconds], [and]** 48 **[thirds]. In the case of Mars, [it is]** 6 **[signs],** 11 **[degrees],** 24 **[minutes],** 9 **[seconds], [and]** 6^2 **[that is, 36 thirds]. In the case of [the *śīghra* of] Mercury, [it is]** 1 **[sign],** 24 **[degrees],** 45 **[minutes],** 18 **[seconds], [and** 0 **thirds]. The multiplier of Jupiter is** 1 **[sign],** 0 **[degrees],** 21 **[minutes],** 6 **[seconds], [and** 0 **thirds]. The multiplier of [the *śīghra* of] Venus is** 7 **[signs],** 15 **[degrees],** 11 **[minutes],** 52 **[seconds], [and]** 48 **[thirds]. For Saturn, [it is]** 0 **[signs],** 12 **[degrees],** 12 **[minutes],** 50 **[seconds], [and]** 24 **[thirds]. In the case of the lunar apogee, [it is]** 1 **[sign],** 10 **[degrees],** 41 **[minutes],** 0 **[seconds], [and]** 54 **[thirds]. For the lunar node, [it is]** 0 **[signs],** 19 **[degrees],** 21 **[minutes],** 11 **[seconds], [and]** 24 **[thirds]. In the case of the Lord of the Year, this [multiplier] in [civil] days and so on is** 1;15,31,31,24. **In the case of the lunar epact, the multiplier in *tithi*s and so on is** 11;3,53,24.

The multipliers are [to be computed as] the revolutions [of the planets in a *kalpa*] divided by the years in a *kalpa*.

The addends [are as follows]. [The addend] of the sun is 6 **[signs],** 0 **[degrees],** 14 **[minutes], [and]** 47 **[seconds]. In the case of the moon, [it is]** 9 **[signs],** 9 **[degrees],** 35 **[minutes], [and]** 42 **[seconds]. In the case of Mars, [it is]** 1 **[signs],** 3 **[degrees],** 34 **[minutes], [and]** 43 **[seconds]. In the case of the *śīghra* of Mercury, [it is]** 4 **[signs],** 0 **[degrees],** 24 **[minutes], [and]** 14 **[seconds]. In the case of Jupiter, [it is]** 2 **[signs],** 14 **[degrees],** 16 **[minutes], [and]** 12 **seconds. In the case of the *śīghra* of Venus, [it is]** 10 **[signs],** 4 **[degrees],** 35 **[minutes], [and]** 30 **[seconds]. In the case of Saturn, [it is]** 2 **[signs],** 19 **[degrees],** 22 **[minutes], [and]** 17 **[seconds]. In the case of the [lunar] apogee, [it is]** 7 **[signs],** 7 **[degrees],** 35 **[minutes], [and]** 14 **[seconds]. In the case of the [lunar] node, [it is]** 0 **[signs],** 11 **[degrees],** 39 **[minutes], [and]** 53 **[seconds]. In the case of the Lord of the Year, [it is]** 4;18,53,0 **in [civil] days and so on. In the case of the epact, [it is]** 2;29,34,0 **in *tithi*s and so on.**

In these verses, Jñānarāja gives the multipliers and addends for the planets, the lunar apogee, the lunar node, the Lord of the Year, and the epact. The epact was defined earlier (verses 54–55). The multiplier for the epact is the epact accumulated during a *saura* year, while the addend for the epact is the accumulated epact at the beginning of *śaka* 1425. The Lord of the Year for a given year is the time in days between the sun's entry into Aries (which marks the beginning of the year) and the end of the preceding week. As such, it determines the weekday that begins the current year, and it is called "Lord of the Year" because the god associated with this weekday is said to rule that year. The addend for the Lord of the Year is the Lord of the Year for *śaka* 1425, and its multiplier is the excess of an integral number of weeks accumulated during a *saura* year. Note that for the epact and the Lord of the Year, the addends correspond to the beginning of *śaka* 1425 rather than to the epoch later in that year. Why this is so is clear from the definitions of the two.

The multipliers and the addends are listed in Table 4. In

Planet	Multiplier	Addends
The sun	—	$6^s0°14'47''$
The moon	$4^s12°46'40''48'''$	$9^s9°35'42''$
Mars	$6^s11°24'9''36'''$	$1^s3°34'43''$ $1^s3°42'35''$
Mercury (*śīghra*)	$1^s24°45'18''0'''$	$4^s0°24'14''$ $4^s0°25'20''$
Jupiter	$1^s0°21'6''0'''$	$2^s14°16'12''$
Venus (*śīghra*)	$7^s15°11'52''48'''$	$10^s4°35'30''$ $10^s4°35'29''$
Saturn	$0^s12°12'50''24'''$	$2^s19°22'17''$
Lunar apogee	$1^s10°41'0''54'''$	$7^s7°35'14''$
Lunar node	$0^s19°21'11''24'''$	$0^s11°39'53''$ $0^s11°40'9''$
Lord of the Year	1;15,31,31,24	4;18,53,0 4;18,53,25,36
Epact	11;3,53,24,0	2;29,34,0 2;29,33,36

Table 4: Multipliers and addends

some cases, two values are given for the addend. If so, the first value is the one given by Jñānarāja, whereas the second is a direct computation. If there is only one value given, there is no difference between the value given by Jñānarāja and the one found through direct computation.

The multiplier of a planet is the angular distance traveled by the mean planet during a *saura* year. If R_p is the number of revolutions of the planet in a *mahāyuga*, it moves a total of $360 \cdot R_p$ degrees during a *mahāyuga*.[185] Therefore, the number of degrees that the planet moves during a *saura* year is $\frac{360 \cdot R_p}{Y}$. Depending on the planet, this number might exceed 360 degrees. If so, we diminish the result by the largest multiple of 360 contained in it, which leaves us with a number g satisfying $0 \leq g < 360$. For example, if the planet travels $1^{\text{rot}}2^s29°1'23''12'''$ in a *saura* year, we get $g = 2^s29°1'23''12'''$ (1^{rot} is one rotation, that is, 360°, and 1^s is a sign, that is, 30°). This number g is the planet's multiplier. It gives the degrees in excess of a whole number of rotations that the planet moves during a *saura* year.

Since $\frac{360 \cdot 60^3}{Y} = 18$, any given multiplier for a planet (or for the moon's apogee and node) will have at most three sexagesimal places. In fact, all the planetary multipliers as given by Jñānarāja are exact. Similarly, the values of the Lord of the Year and the epact are both exact.

Now, the addends are mean planetary positions (as well as the accumulated Lord of the Year and epact) corresponding to a given point in time, the epoch, as explained in the commentary to verse 57. We know that the epoch falls in the year *śaka* 1425. The addend of the sun, which is the mean longitude of the sun at the epoch, will help us to determine it precisely. Since the addend of the sun is $6^s0°14'47''$, the mean sun is $0°14'47''$ into the sign Libra at the time of the epoch, or, equivalently, $\frac{6;0,14,47}{12}$ of a *saura* year (based on mean motion) has elapsed between the beginning of the year *śaka* 1425 and the epoch. This is slightly more than half a *saura* year.

From verses 4–5, we know that 1,955,883,179 *saura* years elapsed between the commencement of planetary motion and the

[185] While Jñānarāja says that the multiplier of a planet is to be found as the revolutions of the planet in a *kalpa* divided by the *saura* years in a *kalpa*, the revolutions and *saura* years in a *mahāyuga* will do just as well. Also, more precisely, the ratio has to be multiplied by 360, as we want the multiplier expressed in degrees, not in rotations.

commencement of the *śaka* era. Therefore,

$$1{,}955{,}883{,}179 + 1{,}425 = 1{,}955{,}884{,}604 \quad (33)$$

saura years elapsed between the commencement of planetary motion and the beginning of the year *śaka* 1425. If $R_☾$ denote the revolutions of the moon in a *mahāyuga*, the longitude of the mean moon is therefore

$$1{,}955{,}884{,}604 \cdot \frac{R_☾}{Y} = 29^\circ 54' 43'' 12''' \quad (34)$$

at the beginning of *śaka* 1425 (the sun's mean longitude at the beginning of *śaka* 1425 is, of course, $0^s 0^\circ 0' 0'' 0'''$). Furthermore, since the mean moon moves

$$\frac{1}{2} \cdot \frac{360 \cdot R_☾}{Y} = 6^{\text{rot}} 8^s 6^\circ 23' 20'' 24''' \quad (35)$$

during half of a *saura* year, it is easy to see that the mean moon will overtake the mean sun 6 times between the beginning of *śaka* 1425 and the epoch. In Jñānarāja's system, a month begins at a conjunction of the sun and the moon, and the beginning of a *saura* year takes place in the month of *caitra*. Hence, the month current at the time of the epoch is *āśvina*. Furthermore, noting that the differences between the mean longitudes of the moon and the sun at the time of the epoch are

$$9^s 9^\circ 35' 42'' - 6^s 0^\circ 14' 47'' = 3^s 9^\circ 20' 55'' = (8 \cdot 12 + 3)^\circ 20' 55'' \quad (36)$$

and

$$\frac{3^\circ 20' 55''}{12} = 0;16,44,35, \quad (37)$$

we see that about 8;16,45 *tithi*s have elapsed of the month of *āśvina* at the time of the epoch.

A computation of mean longitudes based on the formulae of the *Sūryasiddhānta* shows that the epoch corresponds to local sunrise in Jñānarāja's area on Friday, September 29, 1503 CE.[186]

Since 1,955,884,604 years have elapsed between the beginning of planetary motion and the beginning of the year *śaka*

[186] I am grateful to Michio Yano for running this computation for me.

1425, the accumulated Lord of the Year for that time can be found as

$$\begin{aligned} 1{,}955{,}884{,}604 \cdot 1;15{,}31{,}31{,}24 &= \frac{664{,}735{,}254{,}025{,}007}{270{,}000} \\ &\equiv 4;18{,}53{,}25{,}36 \bmod 7 \quad (38) \end{aligned}$$

and the accumulated epact as

$$\begin{aligned} 1{,}955{,}884{,}604 \cdot 11;3{,}53{,}24{,}0 &= \frac{32{,}462{,}305{,}743{,}739}{1{,}500} \\ &\equiv 2;29{,}33{,}36 \bmod 30. \quad (39) \end{aligned}$$

(65) **The day count [for the beginning of the year] is the *tithi*s elapsed since the beginning of *caitra* diminished [first] by *tithi*s of the epact and [then] by the *ghaṭikā*s of the Lord of the Year. The weekday is [counted] from the Lord of the Year. The day count is to be imagined as follows: the difference between half a *saura* year and the day count for the sake of the computation of [the mean longitudes of] the planets.**

The day count is here defined as the elapsed *tithi*s since the beginning of the month of *caitra* diminished by two quantities: the accumulated epact and the *ghaṭikā*s, that is, the fractional part, of the Lord of the Year. By subtracting the accumulated epact from the elapsed *tithi*s, we arrive at the time of the sun's entry into Aries, which is the beginning of the year. However, the sun does not necessarily enter Aries right at the beginning of a weekday. By the definition of the Lord of the Year, if we subtract the *ghaṭikā*s of the Lord of the Year from the previous result, we get the point in time of the beginning of the weekday during which the sun's entry into Aries occurred. In the case of the year *śaka* 1425, the sun's entry into Aries occured 4;18,53 civil days after the beginning of the previous Sunday. Subtracting the *ghaṭikā*s of the Lord of the Year, that is, 18;53 *ghaṭikā*s, from the time of the sun's entry into Aries, we get the beginning of the weekday during which the sun entered Aries in *śaka* 1425.

Although it is not made clear, it appears that what Jñānarāja intends to do with this verse is to establish a day count starting from the epochal date. Assuming that more than half of *śaka*

1425 has passed, if the day count established for the beginning of the year is subtracted from half a year, we get a new day count starting from the epochal date. This is peculiar, though, as the epochal date corresponds to local sunrise, whereas the *saurapakṣa* operates with a midnight system.

Note that what is translated merely as "epact" in the verse is really given as *vigatartuśuddhi*. The word *vigatartu* is literally "elapsed season," but it is not clear how exactly it is to be interpreted here.

(66) **The degrees of [the mean longitude of] the sun [are found as follows]. The day count is diminished by its own 60th part, [and then the result is] increased in its first sexagesimal place by its own seventh part and decreased in its second sexagesimal place by one-fourth of the day count.**

The result is a negative and positive contribution to the planets depending on the hemisphere that the sun [is in].

This and the following seven verses give formulae for computing the mean longitudes of the planets from the day count. Note that in order for the formulae to work properly, the day count in question has to start from the commencement of planetary motion (or at least from a point in time where the mean planets are all at Aries 0°).

The last part of the verse seems out of place here. What is translated in that part as "result" is the Sanskrit word *phala*, which could also be taken as "equation of center." At any rate, this passage does not make any sense where it is found.

Now, if the day count is a, then the angular distance in degrees traveled by a planet during the day count s is

$$s = a \cdot \frac{360 \cdot R_p}{C}, \tag{40}$$

where R_p is the number of revolutions of the planet in a *mahāyuga*. Clearly, $\bar{\lambda}_p \equiv s \bmod 360$, where $\bar{\lambda}_p$ is the longitude of the mean planet.

Let $\bar{\lambda}_{\odot}$ be the mean longitude of the sun and $R_{\odot}$ be the revolutions of the sun in a *mahāyuga*. Then according to the verse we have

$$\bar{\lambda}_{\odot} \equiv a \cdot \left(1 - \frac{1}{60} + \left(1 - \frac{1}{60}\right) \cdot \frac{1}{7 \cdot 60} - \frac{1}{4 \cdot 60^2}\right) \bmod 360, \quad (41)$$

which means that

$$\bar{\lambda}_{\odot} \equiv \frac{99{,}349}{100{,}800} \cdot a \bmod 360. \quad (42)$$

From this, the mean velocity of the sun during one civil day comes out to be $\frac{99{,}349}{100{,}800} = 0;59,8,10,42,51°$. Since

$$\begin{aligned} \frac{99{,}349}{100{,}800} - \frac{360 \cdot R_{\odot}}{C} &= \frac{99{,}573{,}493}{39{,}763{,}529{,}265{,}600} \\ &< 0;0,0,0,33, \end{aligned} \quad (43)$$

we see that this is a good approximation.

It should be noted that the large fraction computed above, namely, $\frac{99{,}573{,}493}{39{,}763{,}529{,}265{,}600}$, was likely never of interest to Jñānarāja, but is included here to show the accuracy of the approximation. Similar fractions are included in the following seven verses as well for the same reason.

It is likely that Jñānarāja derived this and the following seven approximations through knowing the sexagesimal value of $\frac{360 \cdot R_p}{C}$, which he tweaked to get an approximation. As is seen here and in the following verses, the approximations still show their sexagesimal origins.

(67) **The day count is multiplied by 13 and put down in two places. [One place is] multiplied by 10 and divided by 737, [and then] added to the result [in the other place]. [The result of] that with its second sexagesimal place diminished by the eighth part of the day count is [the mean longitude of] the moon in degrees and so on.**

When Jñānarāja says that a quantity, say, x, is to be subtracted from the second sexagesimal place of another quantity, say, y, what is intended is $y - \frac{x}{60^2}$.

Let $\bar{\lambda}_{☾}$ be the mean longitude of the moon and $R_{☾}$ be the number of revolutions of the moon in a *mahāyuga*. The formula given for the mean longitude of the moon is

$$\bar{\lambda}_{☾} \equiv a \cdot \left(13 + 13 \cdot \frac{10}{737} - \frac{1}{8 \cdot 60^2}\right) \bmod 360. \quad (44)$$

In other words,

$$\bar{\lambda}_{☾} \equiv a \cdot \frac{279{,}676{,}063}{21{,}225{,}600} \bmod 360, \tag{45}$$

which gives 13;10,34,52,54,25,24° as the mean velocity of the moon during one civil day.

Since

$$\begin{aligned}\left(13 + 13 \cdot \frac{10}{737} - \frac{1}{8 \cdot 60^2}\right) &- \frac{360 \cdot R_{☾}}{C} \\ &= \frac{32{,}693{,}993{,}791}{8{,}373{,}063{,}162{,}499{,}200} \\ &< 0;0,0,0,51, \end{aligned} \tag{46}$$

this is a good approximation.

(68) **Half of the day count [is] increased by its own 21st part. [Then] the day count diminished by a quarter [of itself] is added to the second sexagesimal place [of the result]. [This is] the [mean longitude of] Mars in degrees and so on.**

Let $\bar{\lambda}_{♂}$ be the mean longitude of Mars and $R_{♂}$ be the number of revolutions of Mars in a *mahāyuga*. The formula given is

$$\bar{\lambda}_{♂} \equiv a \cdot \left(\frac{1}{2} + \frac{1}{2 \cdot 21} + \frac{3}{4 \cdot 60^2}\right) \bmod 360, \tag{47}$$

or

$$\bar{\lambda}_{♂} \equiv \frac{5{,}869}{11{,}200} \cdot a \bmod 360. \tag{48}$$

Since

$$\begin{aligned}\frac{360 \cdot R_{♂}}{C} - \left(\frac{1}{2} + \frac{1}{2 \cdot 21} + \frac{3}{4 \cdot 60^2}\right) &= \frac{6{,}722{,}867}{4{,}418{,}169{,}918{,}400} \\ &< 0;0,0,0,20, \end{aligned} \tag{49}$$

this is again a good approximation.

(69) **[To compute] the degrees of [the mean longitude of] the *śīghra* apogee of Mercury, the day count is multiplied by 4, [and the result is first] increased in its first sexagesimal place by what is obtained [as the quotient**

of the division of] 9 into the day count multiplied by 50, [and then] diminished by the [day] count in the second sexagesimal place.

Let $\bar{\lambda}_{☿}$ be the mean longitude of the *śīghra* apogee of Mercury and $R_{☿}$ be the number of revolutions of the *śīghra* apogee of Mercury in a *mahāyuga*. The verse gives the formula

$$\bar{\lambda}_{☿} \equiv a \cdot \left(4 + \frac{50}{9 \cdot 60} - \frac{1}{60^2}\right) \bmod 360, \tag{50}$$

which is equivalent to

$$\bar{\lambda}_{☿} \equiv a \cdot \frac{44{,}197}{10{,}800} \bmod 360. \tag{51}$$

Since

$$\frac{360 \cdot R_{☿}}{C} - \left(4 + \frac{50}{9 \cdot 60} - \frac{1}{60^2}\right) < 0;0,0,0,41, \tag{52}$$

this is a good approximation.

(70) **In the case of [the mean longitude of] Jupiter, [the procedure is as follows]. [First] the day count is multiplied by 3 and divided by 20, and [the result] is subtracted from the day count. [This quantity is applied] negatively [to] the second sexagesimal place [of] the minutes of arc and so on that are given by the day count multiplied by 5.**

Let $\bar{\lambda}_{♃}$ be the mean longitude of Jupiter and $R_{♃}$ be the revolutions of Jupiter in a *mahāyuga*. The formula given for the mean longitude of Jupiter is

$$\bar{\lambda}_{♃} \equiv a \cdot \left(\frac{5}{60} - \frac{1 - \frac{3}{20}}{60^2}\right) \bmod 360, \tag{53}$$

that is,

$$\bar{\lambda}_{♃} \equiv a \cdot \frac{5{,}983}{72{,}000} \bmod 360. \tag{54}$$

Since

$$\begin{aligned}\left(\frac{5}{60} - \frac{1 - \frac{3}{20}}{60^2}\right) - \frac{360 \cdot R_{♃}}{C} &= \frac{24{,}991{,}231}{28{,}402{,}520{,}904{,}000} \\ &< 0;0,0,0,12,\end{aligned} \tag{55}$$

we again have a good approximation.

(71[a–b]) **In the case of [the mean longitude of] the *śīghra* [apogee] of Venus, the result for the sun increased by its own half is [further] increased by the eighth part of the [day] count diminished by [its own] 100th part.**

Let $\bar{\lambda}_{♀}$ be the mean longitude of the *śīghra* apogee of Venus and $R_{♀}$ be the number of revolutions of the *śīghra* apogee of Venus in a *mahāyuga*. The formula for the mean longitude of the *śīghra* apogee of Venus is

$$\begin{aligned}\bar{\lambda}_{♀} &\equiv a \cdot \left(1 - \frac{1}{60} + \left(1 - \frac{1}{60}\right) \cdot \frac{1}{7 \cdot 60} - \frac{1}{4 \cdot 60^2}\right) \cdot \frac{3}{2} \\ &\quad + a \cdot \left(\frac{1}{8} - \frac{1}{8 \cdot 100}\right) \bmod 360, \qquad (56)\end{aligned}$$

which means that

$$\bar{\lambda}_{♀} \equiv \frac{21{,}533}{13{,}440} \cdot a \bmod 360. \qquad (57)$$

Since

$$\frac{21{,}533}{13{,}440} - \frac{360 \cdot R_{♀}}{C} = \frac{60{,}137{,}981}{5{,}301{,}803{,}902{,}080} < 0;0,0,3, \qquad (58)$$

this is a good approximation.

(71[c–d]) **[The mean longitude of] Saturn in degrees and so on is the day count divided by 30 with its first sexagesimal place increased by the 100th and 60th part of the [day] count.**

Let $\bar{\lambda}_{♄}$ be the mean longitude of Saturn and $R_{♄}$ be the number of revolutions of Saturn in a *mahāyuga*. For the mean longitude of Saturn, we are given the formula

$$\bar{\lambda}_{♄} \equiv a \cdot \left(\frac{1}{30} + \frac{1}{60 \cdot 160}\right) \bmod 360, \qquad (59)$$

that is,

$$\bar{\lambda}_{♄} \equiv \frac{107}{3{,}200} \cdot a \bmod 360. \qquad (60)$$

Since

$$\frac{360 \cdot R_{♄}}{C} - \left(\frac{1}{30} + \frac{1}{60 \cdot 160}\right) = \frac{2{,}282{,}101}{1{,}262{,}334{,}262{,}400} < 0;0,0,0,24, \quad (61)$$

this is a good approximation.

(72[a–c]) **[The longitude of] the lunar apogee in degrees and so on [can be computed as follows]: the day count is divided by 10. In the first sexagesimal place [the result] is increased by the day count diminished by its own third part and increased by the 41st part [of the day count diminished by its own third part].**

Let λ_A be the longitude of the lunar apogee and R_A be the number of revolutions of the lunar apogee. The longitude of the lunar apogee can be found from the formula

$$\lambda_A \equiv a \cdot \left(\frac{1}{10} + \frac{1}{60} \cdot \left(\left(1 - \frac{1}{3}\right) + \frac{1}{41} \cdot \left(1 - \frac{1}{3}\right)\right)\right) \bmod 360, \quad (62)$$

which means that

$$\lambda_A \equiv a \cdot \frac{137}{1{,}230} \bmod 360. \quad (63)$$

Since

$$\frac{360 \cdot R_A}{C} - \frac{137}{1{,}230} = \frac{386{,}491}{485{,}209{,}732{,}110} < 0;0,0,0,11, \quad (64)$$

it is a good approximation.

(72[d]–73) **[Now,] according to the regular order, I will explain the [formula for] the lunar node. The day count [is first] multiplied by 10 and divided by 3, [and then that quantity is] diminished by its own 22nd part. [The result is the mean longitude of] the lunar node in minutes of arc and so on.**

[The mean longitudes of] the planets are [equal to] their own *dhruva*s at [sun]rise.

Let λ_{Ω} be the longitude of the lunar node and R_{Ω} be the number of revolutions of the lunar node in a *mahāyuga*. The longitude of the lunar node can be found from the formula

$$\lambda_{\Omega} \equiv a \cdot \left(\frac{10}{3} - \frac{1}{22} \cdot \frac{10}{3}\right) \cdot \frac{1}{60} \bmod 360. \tag{65}$$

Therefore,

$$\lambda_{\Omega} \equiv a \cdot \frac{7}{132} \bmod 360. \tag{66}$$

Since

$$\begin{aligned} \left(\frac{10}{3} - \frac{1}{22} \cdot \frac{10}{3}\right) \cdot \frac{1}{60} - \frac{360 \cdot R_{\Omega}}{C} &= \frac{2{,}368{,}759}{52{,}071{,}288{,}324} \\ &< 0;0,0,10, \end{aligned} \tag{67}$$

it is a good approximation.

The last part of the verse is not clear.

(74) **The [mean] velocity of a planet [which has a mean longitude] that is not known is multiplied by the known [mean longitude of another] planet and divided by the [mean] velocity of the planet [that has a mean longitude] that is known. The result is the [mean longitude of the] planet [that has a mean longitude] that was not known [previously].**

The above attempt at a literal translation of the verse does not read well, but the idea is as follows. Given two planets, say, P_1 and P_2, let their mean longitudes and mean velocities be $\bar{\lambda}_1$ and $\bar{v}_1$, and $\bar{\lambda}_2$ and $\bar{v}_2$, respectively. Assuming that $\bar{\lambda}_1$, $\bar{v}_1$, and $\bar{v}_2$ are known, the verse states that $\bar{\lambda}_2$ can be found as

$$\bar{\lambda}_2 = \frac{\bar{v}_2 \cdot \bar{\lambda}_1}{\bar{v}_1}. \tag{68}$$

As can be easily seen, this formula does not work. For example, in verses 58–64, we had a situation where the mean longitude of the sun is $\bar{\lambda}_{\odot} = 180°14'47''$ and the mean longitude of the moon is $\bar{\lambda}_{☾} = 279°35'42''$. Assuming that the sun is the planet whose mean longitude is known, the formula gives us, incorrectly, that

$$\bar{\lambda}_{☾} = \frac{\bar{v}_{☾} \cdot \bar{\lambda}_{\odot}}{\bar{v}_{\odot}} = \frac{13;10,34 \cdot 180;14,47}{0;59,8} \equiv 249;45,14 \bmod 360. \tag{69}$$

The reason for the failure of the formula in the above example is that we have not taken into account that each planet has moved a certain number of whole rotations since the commencement of planetary motion. If, instead of longitudes between 0° and 360°, we use the total angular distance traveled by the planet since the commencement of planetary motion, the formula works.

(75) **If an unattained intercalary month is attained, then the epact is diminished by 30 [and] the day count is to be computed from that at the middle of the year. Then the [number of] elapsed intercalary months is diminished by 1.**

The idea is as follows. Gradually, over time, a residue of an intercalary month accumulates. When this residue reaches 1, it contains a full intercalary month, which is then inserted. It is not clear why the number of intercalary months should be diminished by 1, though.

(76) **A normal lunar month [occurs] when it contains a *saṅkrānti*. When this is not the case, an omitted [month] or an intercalary [month] is produced.**

A *saṅkrānti* is the passage of the sun from one zodiacal sign to the next. If such a passage occurs during a lunar month, that month is said to be a regular month, or a normal month. The majority of months fall in this category. If no *saṅkrānti* occurs during a month, that month is called an intercalary month. If two *saṅkrānti*s occur during a month, that month is called an omitted month.

(77) **The lunar month during which the entry of the sun into Aries takes place is called *madhu*. [The months] *mādhava* and so on occur with [the entry of the sun into] Taurus and so on.**

The first month of the year, *caitra* or *madhu*, is the lunar month during which the sun enters the sign Aries. Similarly, during the second month the sun enters Taurus, and so on. The names of the months are, in order, *madhu*, *mādhava*, *śukra*, *śuci*, *nabha*, *nabhasya*, *iṣa*, *ūrja*, *saha*, *sahasya*, *tapa*, and *tapasya*. In addition to these descriptive names, the months are also named after the *nakṣatra*, or constellation in the path of the moon, at which the full moon in that month occurs. In this system, the names of the months are *caitra*, *vaiśākha*, *jyaiṣṭha*, *āṣāḍha*, *śrāvaṇa*, *bhādrapada*, *āśvina*, *kārttika*, *mārga-śīrṣa*, *pauṣya*, *māgha*, and *phālguna*.

(78) **Since a *saura* year is [measured] by 12 lunar months, 11 *tithi*s, 3 *ghaṭikā*s, and 53 *pala*s, whatever is greater than the increase of the year, that is an intercalary month [measured] by 32;16. 30 *tithi*s is a mean [intercalary month].**

The first part is clear, for

$$\frac{T}{Y} - 12 \cdot 30 = \frac{66{,}389}{6{,}000} = 11;3,53,24. \qquad (70)$$

We thus have the duration of a *saura* year measured in *tithi*s. However, the second part of the verse is not clear (but compare the next verse).

(79) **If one civil day [comes about] by means of the mean motion of the sun, then what [arises] in a sign? A *saura* month, and the lunar month is [found] in the difference of the velocities. If there is one civil day of the sun, then what is in a revolution? A *saura* [month] is greater than a lunar month, and an intercalary month is [measured] by 32;16.**

The verse is not clear (but compare the previous verse).

(80) **A month [begins] from a conjunction [of the sun and the moon] and ends at the [next] conjunction. If there is an entry of the sun into a sign during that [time], [the month] is a regular one; otherwise, it is an intercalary one.**

The significance of the word *maṇḍalanāḍikāntaṃ* is not clear to me and is not included in the translation. Otherwise, the verse is clear. A month is the duration from one conjunction of the sun and the moon to the next. Normally the sun will enter a sign in that time, in which case the month is a regular month. However, if this is not the case, the month is either omitted or intercalary; see verse 76. Here only the intercalary option is given.

(81–82) **Whatever has been stated [in the preceding] with respect to those at the center of the earth I will now explain with respect to those on the surface of the earth.**

The computed time of conjunction [of the sun and the moon] is corrected [when] rectified by the longitudinal parallax.

Since at the corrected time of conjunction [of the sun and the moon], the sun and the moon are certainly situated on the [same] line of vision, those who know the sphere say that.

Note that we have to imagine that there is an "observer" or "observers" at the earth's center. If such an observer is postulated, he will see the conjunction of the sun and the moon at a different time than an observer on the surface of the earth. This is due to parallax, a topic that is dealt with exhaustively in *Siddhāntasundara* 2.6.1–5.

(83–84) **Whatever has elapsed of the *kali*[*yuga*] is divided by 180,000. The smaller of what has elapsed and what is to come of it [*kaliyuga*] is multiplied [separately] by 1, 2, 3, 24, 22, 30, 90, and 7 [the result corresponding to the seven planets].**

[The longitudes of] the sun, Saturn, and Mars are increased by [the result expressed in] degrees divided by 90,000, and [the longitudes of] the others [that is, the moon, Mercury, Venus, and Jupiter] are diminished [by it].

The intelligent Dāmodara says that in this way, by means of this [correction], there is true identity with what is observed for the planets.

Jñānarāja here follows a set of *bīja* (literally, seed) correc-

Planet	Correction (*bīja*)
The sun	1
Saturn	2
Mars	3
The moon	−24
Mercury	−22
Venus	−30
Jupiter	−90
The moon's apogee?	−7

Table 5: Table of *bīja*s

tions given by Dāmodara.[187] Unfortunately, Dāmodara's work has not been published or studied.

A *bīja* correction often converts from data of one *pakṣa*, or astronomical school, to another, but it is equally often obscure. The *bīja*s of Dāmodara given here by Jñānarāja are enigmatic, and it is not clear why Jñānarāja chose to include them.

It is not entirely clear what number corresponds to what planet. First of all, there are eight numbers, but only seven planets. The remaining number might belong to the moon's apogee, but that leaves the moon's node without a *bīja* correction. Secondly, the order is not clear. If it is assumed that the first three numbers belong to the sun, Saturn, and Mars, respectively, whereas the next five belong to the moon, Mercury, Venus, Jupiter, and the moon's apogee, Table 5 gives the *bīja*s with their sign. It should be noted, however, that is possible that the planets are meant to be taken in their usual order (the sun, the moon, Mars, Mercury, Jupiter, Venus, Saturn).

(85–86) **Kanyā is located** 125 ***yojana*s from Laṅkā [to the north]. Kāntī is [north of Kanyā] by** 32 **[*yojana*s]. Svāmī is [north of Kāntī] by** 80 **[*yojana*s]. Sagara is [north of Svāmī] by** 20 **[*yojana*s]. Mallāri is [north of Sagara] by** 15 **[*yojana*s]. Paryalī is north [of Mallāri] by** 8 ***yojana*s. The city [of] Vatsagulma is [north of Paryalī] by** 10 **[*yojana*s]. The city of Ujjayinī is [north of Vatsagulma] by**

[187]For Dāmodara, see Dikshit 1969–81, 2.125–127.

50 [*yojanas*]. **Kurukṣetra is [north of] that [place] by** 110 [*yojanas*]. **[The mountain] Meru is [north of] that [place] by** 825 ***yojanas*. In this way the prime meridian of the earth is explained.**

In Indian astronomy, the prime meridian passes through Ujjayinī (the modern Ujjain in Madhya Pradesh). In other words, the prime meridian is the great circle from the North Pole to the South Pole that passes through Ujjayinī. These two verses list a number of locations on the prime meridian, as well as the distance in *yojanas* between two successive locations. The list starts from Laṅkā in the south and proceeds to Meru in the north. Laṅkā is on the terrestrial equator and Meru is on the North Pole, so the locations mentioned span exactly one-quarter of the earth's circumference.

The circumference of the earth is 5,059 *yojanas*,[188] and thus one-quarter of the circumference is $\frac{5{,}059}{4} = 1{,}264\frac{3}{4}$ *yojanas*. However, if we add up the distances given, we get

$$125 + 32 + 80 + 20 + 15 + 8 + 10 + 50 + 110 + 825 = 1{,}275 \quad (71)$$

yojanas. We use a rule given in the section on three questions,[189] according to which the difference between the latitudes of two locations on the same meridian is to be multiplied by 14 to get the *yojanas* between the locations. Using this rule to find the difference in latitude between two locations 1,275 *yojanas* apart on the same meridian, we get

$$\frac{1{,}275}{14} = 90\frac{1}{14} \approx 90;4, \quad (72)$$

which is fairly close to the expected 90°. Cintāmaṇi makes reference to this rule in his commentary, saying that it can be used to compute the distance between locations on the prime meridian and giving one example.

In some manuscripts, Kanyā is called Devakanyā, and Kāntī and the distance of 32 *yojanas* between Kāntī and Kanyā are omitted. This gives a total of $1{,}275 - 32 = 1{,}243$ *yojanas*, which is less accurate than the previous total.

[188]See *Siddhāntasundara* 1.1.74.
[189]See *Siddhāntasundara* 2.3.18.

(87) **The daily motion of the planets is multiplied by the *yojanas* between a city on the prime meridian and the given city, [the two being on the same latitude circle], and divided by the corrected circumference of the earth. The minutes of arc of the result are applied positively or negatively to the mean planet according to whether the given city is located west or east [of the prime meridian].**

As we have seen, the planetary positions computed for a given day correspond to midnight at Laṅkā. If we are on the prime meridian, they will correspond to our midnight as well. However, if we are not on the prime meridian, the positions need to be adjusted in order to correspond to our local midnight. The correction that is carried out to achieve this is called the longitudinal correction (*deśāntara*).

Suppose that the given location, P_1, is d *yojanas* from a location on the prime meridian, P_2, along a latitude circle (a circle through all locations with the same latitude). If we are not on the terrestrial equator, the circumference of our latitude circle will be smaller than the circumference of the earth. The circumference of our latitude circle is called the corrected circumference of the earth, and how it is computed will be explained in the next verse. For now, assume that its circumference is c'. Like the terrestrial equator, the corrected circumference can be thought of as measuring time, its whole circumference corresponding to 60 *ghaṭikās*, that is, a nychthemeron.

Let $\overline{v}$ be the mean velocity of a given planet. The quantity

$$\overline{v} \cdot \frac{d}{c'} \tag{73}$$

is the distance traveled by the mean planet in the time between midnight on the prime meridian and midnight at the given location. Since the mean velocity of a planet is generally given in minutes of arc per civil day, the result will be in minutes of arc. These minutes of arc are added to the mean longitude of the planet if the given location is west of the prime meridian, because midnight will occur later here than on the prime meridian. Similarly, the minutes of arc are subtracted from the mean longitude of the planet if the given location is east of the prime meridian.

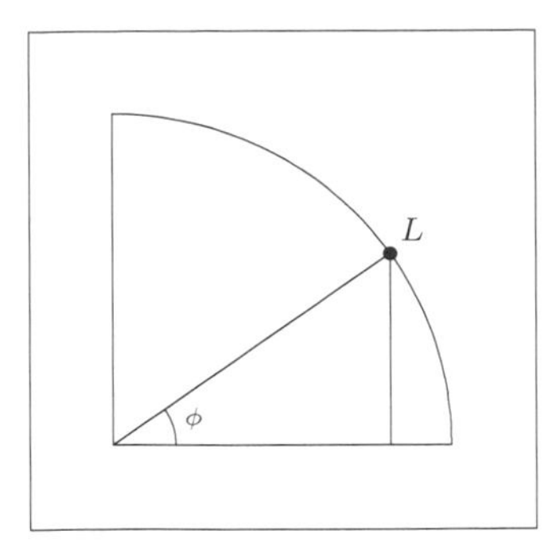

Figure 7: The corrected circumference of the earth for the latitude ϕ

(88) **The corrected circumference on the earth is computed by means of the *yojanas* between the given city and the mountain of the gods.**

The circumference of the earth multiplied by the Sine of the co-latitude and divided by the radius is said to be its [the corrected circumference's] measure.

Let c be the circumference of the earth and c' the corrected circumference corresponding to the latitude ϕ. Then

$$c' = c \cdot \frac{\mathrm{Sin}(\bar{\phi})}{R}. \tag{74}$$

That this formula is correct is easily seen from Figure 7. The figure shows a section of the sphere of the earth, the horizontal line being the terrestrial equator. The given location, L, has the latitude ϕ. Let r be the radius of the earth and r' the radius corresponding to the corrected circumference. It is clear from the figure that

$$r' = r \cdot \cos(\phi) = r \cdot \sin(\bar{\phi}) = r \cdot \frac{\mathrm{Sin}(\bar{\phi})}{R}. \tag{75}$$

Since $\frac{r'}{r} = \frac{c'}{c}$, we get the formula of the verse:

$$c' = c \cdot \frac{\mathrm{Sin}(\bar{\phi})}{R}. \tag{76}$$

(89) That which has been presented by me is very easy and not given by others. Out of a concern of the size in the composing of the *vāsanās*, I condensed it. The wise people who understand the aesthetic abandon what is great and drink the essence, the small circle of the nectar-rayed one [the moon].

(90) [Thus] is presented the planets' mean measure in the beautiful and abundant *tantra* composed by Jñānarāja, the son of Nāganātha, which is the foundation of [any] library.

8. Chapter on Mathematical Astronomy

Section 2: *True Motion*

Having covered mean motion in the previous section, Jñānarāja now turns his attention to true motion. Mean motions only give mean positions of the planets, but in practical reality, we need their true motions. To accomplish this, trigonometry is needed, which is where the section begins.

(1) **Now, when it comes to all auspicious things, [such as] births [of children], performances [of rites], journeys, marriages, and so on, a correct result [arrived at] by means of more correct [positions of the] planets is the most essential thing. Therefore, I am describing this method [for accomplishing this].**

Based on the positions of the planets, auspicious and inauspicious times for various acts and undertakings can be determined. The mean positions described in the previous section are not sufficient for this, and thus the computation of the true positions of the planets is described in this section by Jñānarāja.

(2–5) **The half-Chords, [i.e.,] the Sines, are, in due order:** 225, 449, 671, 890, 1,105, 1,315, 1,520, 1,719, 1,910, 2,093, 2,267, 2,431, 2,585, 2,728, 2,859, 2,978, 3,084, 3,177, 3,256, 3,321, 3,372, 3,409, 3,431, **and** 3,438.

The Versed Sines are equal to the sum of the differences of these taken in the opposite order.

The Chords and Sines (with capitalized first letters) mentioned in the verse will be defined below in the commentary.

In modern mathematics, a sine of an angle is the ratio of the side opposite that angle to the hypotenuse in a right-angled triangle (we will refer to this opposite side as the "upright" in the following). Another way of saying this is that modern sines are normalized with respect to a circle of radius 1. In Indian mathematics, on the other hand, sines are given with respect to circles of different radii. It is most often, as here in the *Siddhāntasundara*, a radius of 3,438, the last value in the list corresponding to the sine of 90°. This value is used by Āryabhaṭa and in the *Sūryasiddhānta*.[190] Other values are used as well; for example, Brahmagupta uses a radius of 3270 in the *Brāhmasphuṭasiddhānta*, Śrīpati uses a radius of 3,415, a value used in the *Vasiṣṭhasiddhānta* as well, and the *Vṛddhavasiṣṭhasiddhānta* has a Sine table with a radius of 1,000.[191]

The choice of the number 3,438 as the radius is in many ways a natural one. Suppose that a circle has the radius R. Then its circumference is $2 \cdot \pi \cdot R$. As there are 21,600 minutes of arc in a circle, the number of minutes of arc per unit of the circumference is $\frac{21{,}600}{2 \cdot \pi \cdot R}$. If $R = 3{,}438$, then $\frac{21600}{2 \cdot \pi \cdot R} \approx 1$. This means that if the angle α is measured in minutes of arc and is small, then $\mathrm{Sin}(\alpha) \approx \alpha$. Using $R = 3{,}438$ can therefore be compared to our modern use of radians instead of degrees.

In the following, the trigonometric radius 3,438 will be called simply "the radius," and it will be denoted by R in formulae. We will write "Sine" when a sine measured with respect to $R = 3{,}438$ is meant, and similarly $\mathrm{Sin}(\alpha)$ will denote the Sine of the angle α. The relationship between this Sine and the modern sine is $\mathrm{Sin}(\alpha) = R \cdot \sin(\alpha)$. Similarly, we will write Cosine, and $\mathrm{Cos}(\alpha)$, Chord and $\mathrm{Crd}(\alpha)$, and Versed Sine and $\mathrm{Vers}(\alpha)$ to denote the cosine, chord, and versed sine functions based on the radius $R = 3{,}438$, respectively. Chords and Versed Sines will be described below and in the commentary to the next verse.

In the Indian tradition, Sines are based on Chords. The Sanskrit words *jīvā* and *jyā* (both literally "bow-string"; a reference to the fact that a chord and its corresponding arc resemble a stringed bow) can mean both "Chord" and "Sine," and in verse

[190] For Āryabhaṭa, see Pingree 1978a, 591, and for the *Sūryasiddhānta*, see *Sūryasiddhānta* 2.15–22.

[191] See *Brāhmasphuṭasiddhānta* 2.2–5; for Brahmagupta's table of Sines, see Yano 1977, 88. For Śrīpati, see Pingree 1978a, 582. For the *Vasiṣṭhasiddhānta*, see *Vasiṣṭhasiddhānta* 38–42, and see also Pingree 1978a, 612. For the *Vṛddhavasiṣṭhasiddhānta*, see *Vṛddhavasiṣṭhasiddhānta* 2.9–10. See also Pingree 1978a, 612.

Sine number	Angle	Sine	Modern value
1	$3°45'$	225	224.85
2	$7°30'$	449	448.74
3	$11°15'$	671	670.72
4	$15°0'$	890	889.81
5	$18°45'$	1,105	1,105.10
6	$22°30'$	1,315	1,315.66
7	$26°15'$	1,520	1,520.58
8	$30°0'$	1,719	1,719.00
9	$33°45'$	1,910	1,910.05
10	$37°30'$	2,093	2,092.92
11	$41°15'$	2,267	2,266.83
12	$45°0'$	2,431	2,431.03
13	$48°45'$	2,585	2,584.82
14	$52°30'$	2,728	2,727.54
15	$56°15'$	2,859	2,858.59
16	$60°0'$	2,978	2,977.39
17	$63°45'$	3,084	3083.44
18	$67°30'$	3177	3,176.29
19	$71°15'$	3,256	3,255.54
20	$75°0'$	3,321	3,320.82
21	$78°45'$	3,372	3,371.93
22	$82°30'$	3,409	3,408.58
23	$86°15'$	3,431	3,430.63
24	$90°0'$	3,438	3,438.00

Table 6: Table of Sines

3, the word *ardhajīvā* is used in the sense of "half-Chord," while in verse 4, the word *jīvā* is used in the sense of "Sine."

Now, the verses give a table of the Sine of each of the 24 multiples of $3°45'$ between $0°$ and $90°$. Note that since $3°45' = 225'$, we can also take it that the table gives us the Sine of each of the 24 multiples of $225'$ between $0° = 0'$ and $90° = 5{,}400'$. Table 6 gives the number of each Sine, the corresponding angle, the Sine of the angle, and the modern value of the Sine for comparison. In the following, we will use the notation Sin_n for $\text{Sin}(n \cdot 3°45')$ for $n = 1, 2, 3, \ldots, 24$, and we will further use Sin_0 for $\text{Sin}(0 \cdot 3°45') = 0$. Sin_n is thus the nth Sine in the table. It will also be convenient to denote the Sine difference $\text{Sin}_n - \text{Sin}_{n-1}$ by ΔSin_n for $n = 1, 2, \ldots, 24$. Similarly, we will also define $\text{Cos}_n = \text{Cos}(n \cdot 3°45')$.

The versed Sine of an angle α is defined as

$$\text{Vers}(\alpha) = R - \text{Cos}(\alpha). \tag{77}$$

The table can be used to compute the versed Sines of the same angles as the Sines by means of the formula

$$\text{Vers}(n \cdot 3°45') = \sum_{k=1}^{n} \Delta\text{Sin}_{25-k} = \sum_{k=1}^{n} (\text{Sin}_{25-k} - \text{Sin}_{24-k}). \tag{78}$$

The formula is straightforward and can be proved as follows:

$$\begin{aligned} \text{Vers}(n \cdot 3°45') &= R - \text{Cos}(n \cdot 3°45') \\ &= R - \text{Sin}(90° - n \cdot 3°45') \\ &= R - \text{Sin}((24 - n) \cdot 3°45') \\ &= \text{Sin}_{24} - \text{Sin}_{24-n} \\ &= \sum_{k=1}^{n} (\text{Sin}_{25-k} - \text{Sin}_{24-k}) \\ &= \sum_{k=1}^{n} \Delta\text{Sin}_{25-k}. \end{aligned} \tag{79}$$

Thus far we have only treated a small number of Sines between 0° and 90°. In the following, we will see how to compute the Sine of an angle that is not necessarily a multiple of 3°45′ and not limited to being between 0° and 90°.

(6) **On a circle, made up of the minutes of arc in a circle and marked with the 96th parts of the circumference, the Chords are to be drawn lying on both of the two points [among the 96 points] that are on a line running east-west. Accordingly, they are 48. Furthermore, their [corresponding] arcs are to be considered; a versed Sine is lying between the arc and [its corresponding] Chord.**

In this and the following verses, Jñānarāja gives a method for how to compute the Sine values given in his table.

The circumference of a circle is divided into 96 equal parts. Two of the points are on the east-west line through the center of

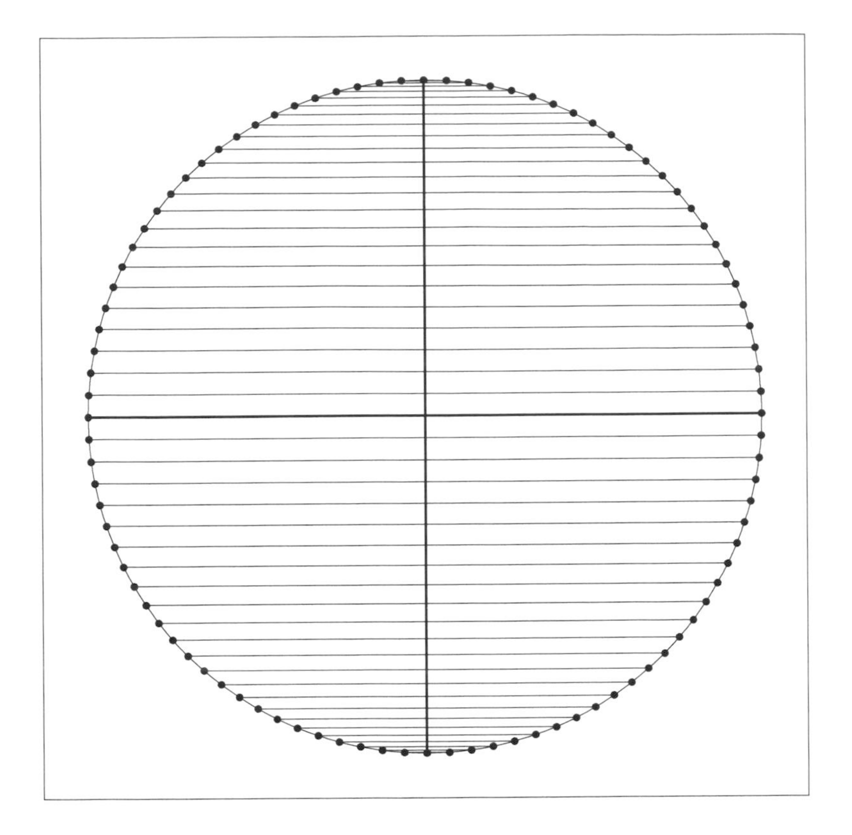

Figure 8: 48 Chords from 96 divisions of the circumference

the circle. The remaining points are connected in pairs such that the line connecting them is parallel to the east-west line. This is shown in Figure 8, where the east-west line is the bold horizontal line. Each horizontal line represents a Chord, namely, the Chord of the angle between and above the two lines. There are, of course, 49 such lines, but two of them are merely points, namely, the lines connecting the north and south points to themselves. As they yield the Chords of $0°$ and $360°$, which subtend the same point on a circle, they are counted as one. This gives a total of 48 Chords.

The definitions of Chords, Sines, and Versed Sines are seen more clearly in Figure 9. The Chord of the arc ACB is the line segment AB, the Sine of $\angle AC$ is the line segment AD, and the Versed Sine of $\angle AC$ is the line segment DC.

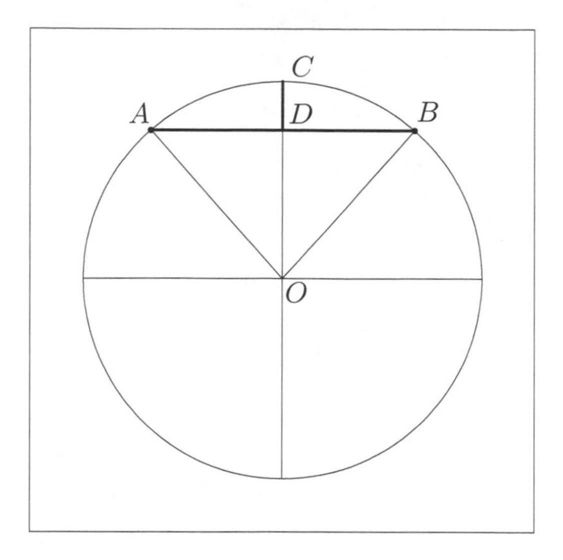

Figure 9: Thc Chord, Sine, and Versed Sine of an angle

(7–8) **The first half[-Chord] is equal to the 96th part of the minutes of arc in a circle [i.e., 225]. [It is] the leg [in a right-angled triangle], and the [corresponding] hypotenuse is the radius. If the upright is found from these two when the hypotenuse [in this way] is the radius, what is it [the upright] when [the hypotenuse] is equal to 225? The result thus [obtained] is an approximate [value of] the difference between the first Sine and the following one.**

[When] the upright is multiplied by 10 and divided by 153, the [Sine] difference is produced. [In other words,] the second [Sine] is produced from the first half[-Chord] increased by that [quantity], and so on like that.

Half of the sum of the traversed [Sine] difference and the current [Sine] difference is the correct Sine difference.

In this way, all the half-Chords are produced in order.

The first Sine, Sin_1, is equal to 225. This is clear from $\text{Sin}_1 = \text{Sin}(3°45') = \text{Sin}(225')$ and from the observation, noted in the commentary on verses 2–5, that $\text{Sin}(\alpha) \approx \alpha$ when α, which is measured in minutes of arc, is small.

The difference between Sin_1 and Sin_2 is then found by means of a proportion. Let a right-angled triangle have as its hypotenuse the radius R, and as its leg $\text{Sin}_1 = 225$. The corresponding upright is, of course, $\text{Cos}_1 = \sqrt{R^2 - \text{Sin}_1^2}$. Now, if a similar triangle has a hypotenuse of length 225, the length of its upright is approximately $\text{Sin}_2 - \text{Sin}_1$. Why is that? The upright in question is of length

$$225 \cdot \frac{\text{Cos}_1}{R} = \frac{\text{Sin}_1}{R} \cdot \text{Cos}_1. \tag{80}$$

That this is approximately equal to $\text{Sin}_2 - \text{Sin}_1$ follows from the following derivation, where we will use the trigonometric relation that $\sin(\alpha) - \sin(\beta) = 2 \cdot \cos(\frac{\alpha+\beta}{2}) \cdot \sin(\frac{\alpha-\beta}{2})$ for all α and β:

$$\begin{aligned} \text{Sin}_2 - \text{Sin}_1 &= \frac{1}{R} \cdot 2 \cdot \text{Cos}\left(\frac{450' + 225'}{2}\right) \cdot \text{Sin}\left(\frac{450' - 225'}{2}\right) \\ &= \frac{2 \cdot \text{Sin}\left(\frac{225'}{2}\right)}{R} \cdot \text{Cos}\left(225' + \frac{225'}{2}\right) \\ &\approx \frac{2 \cdot \frac{225}{2}}{R} \cdot \text{Cos}(225') \\ &= \frac{\text{Sin}_1}{R} \cdot \text{Cos}_1. \end{aligned} \tag{81}$$

Note that since $\text{Cos}(225') = 3{,}430;38$ and $\text{Cos}(225' + \frac{225'}{2}) = 3{,}421;27$, considering these two quantities to be roughly equal is not entirely accurate but acceptable.

If we are looking at the Sine difference $\text{Sin}_{n+1} - \text{Sin}_n$, the upright mentioned is Cos_n. The formula given for finding the Sine differences is

$$\text{Sin}_{n+1} - \text{Sin}_n = \frac{10}{153} \cdot \text{Cos}_n \tag{82}$$

for $n = 1, 2, \ldots, 24$. Noting that

$$\frac{225}{3{,}438} = \frac{10}{153} + \frac{5}{58{,}446} \approx \frac{10}{153}, \tag{83}$$

we can derive the given formula by the same procedure as we just followed in the case $n = 1$. In this case, the derivation becomes

$$\text{Sin}_{n+1} - \text{Sin}_n = \frac{2 \cdot \text{Sin}\left(\frac{225'}{2}\right)}{R} \cdot \text{Cos}\left(n \cdot 225' + \frac{225'}{2}\right)$$

$$\begin{aligned} &\approx \frac{2 \cdot \frac{225}{2}}{R} \cdot \operatorname{Cos}(n \cdot 225') \\ &= \frac{\operatorname{Sin}_1}{R} \cdot \operatorname{Cos}_n. \end{aligned} \tag{84}$$

As before, the weak point of the derivation is the approximation $\operatorname{Cos}(n \cdot 225' + \frac{225'}{2}) \approx \operatorname{Cos}(n \cdot 225')$. However, the approximation works out reasonably in all cases. Jñānarāja presumably derived the identity differently.

The statement that half of the sum of the traversed Sine difference and the current Sine difference is the true Sine difference means the following. If an angle α satisfies $n \cdot 3°45' < \alpha < (n+1) \cdot 3°45'$, then the traversed Sine difference is $\Delta\operatorname{Sin}_n$ and the current Sine difference is $\Delta\operatorname{Sin}_{n+1}$. From the derivation (using the same trigonometric relation as before)

$$\begin{aligned} \frac{\Delta\operatorname{Sin}_{n+1} + \Delta\operatorname{Sin}_n}{2} &= \frac{\operatorname{Sin}_{n+1} - \operatorname{Sin}_{n-1}}{2} \\ &= \frac{1}{R} \cdot \operatorname{Cos}\left(\frac{2 \cdot n \cdot 225'}{2}\right) \cdot \operatorname{Sin}\left(\frac{2 \cdot 255'}{2}\right) \\ &= \frac{\operatorname{Sin}_1}{R} \cdot \operatorname{Cos}_n \\ &\approx \operatorname{Sin}_{n+1} - \operatorname{Sin}_n \\ &= \Delta\operatorname{Sin}_{n+1}, \end{aligned} \tag{85}$$

we get the result stated in the verse. It is, however, an approximation, not an exact result.

(9) **We consider the person who computes all the half-Chords in order from the first half-Chord to be the polestar in the fastening of the motion of the circle of knowers of computation and the circle of stars.**

The polestar remains fixed in the sky while the other stars move. A person who computes the Sines according to Jñānarāja's directions is here compared to the polestar; he fastens the motion of lesser mathematicians.

(10) **There are four quadrants in a circle divided into [12] signs. In the first quadrant, [which spans the three first signs] beginning with Aries, there is an increase in the Sine [when the angle increases]. In the second**

[quadrant], there is a decrease. In the third, there is an increase. And in the fourth, there is a decrease in the leg [i.e., in the Sine].

Like the ecliptic, the circle is here considered to be divided into 12 signs of 30° each, with the sign Aries beginning at 0°. In addition, the circle is divided into four quadrants, the first spanning the first three signs; the second spanning the fourth, fifth, and sixth signs; the third spanning the seventh, eighth, and ninth signs; and the fourth and last spanning the last three signs.

According to Jñānarāja, between 0° and 90° the Sines are increasing. That is, if $0° \leq \alpha < \beta \leq 90°$, then $\mathrm{Sin}(\alpha) < \mathrm{Sin}(\beta)$. Similarly, the Sines are decreasing in the second quadrant, increasing in the third, and finally decreasing in the fourth.

Notice that this differs from the modern sine function, which increases in the first and second quadrants and decreases in the third and fourth quadrants. The reason for this difference is that in the Indian system one always works with positive numbers, and hence every Sine is a positive number. For example, we have $\sin(150°) = -\frac{1}{2}$, but $\mathrm{Sin}(150°) = \frac{1}{2} \cdot 3{,}438 = 1719$.

More than telling us when the Sine function increases and decreases, the verse indirectly tells us how to compute a Sine in the second, third, and fourth quadrants:

$$\mathrm{Sin}(\alpha) = \begin{cases} \mathrm{Sin}(180° - \alpha) & \text{if } 90° < \alpha \leq 180°, \\ \mathrm{Sin}(\alpha - 180°) & \text{if } 180° < \alpha \leq 270°, \text{ and} \\ \mathrm{Sin}(360° - \alpha) & \text{if } 270° < \alpha \leq 360°. \end{cases} \tag{86}$$

In order to be able to compute any given Sine, we therefore only need to be able to compute the Sine of any given angle between 0° and 90°. How to do this will be taken up in the next two verses.

(11) **The complement [of a given arc] is 90 [degrees] diminished by the degrees of the arc. The minutes of arc of those two [quantities] are [separately] divided by 225. The Sine whose number corresponds to [the quotient of] each result is increased by what is attained as a result of the division of the divisor [of the previous division, i.e., 225,] into the remainder [of the previous division]**

multiplied by the difference between the traversed and current Sines.

This verse tells us how to compute the Sine and the Cosine of any given angle between 0° and 90°. The method used is simple linear interpolation.

Let α be an angle measured in minutes of arc and satisfying $0' < \alpha < 5400' = 90°$ ($\mathrm{Sin}(0°)$ and $\mathrm{Sin}(90°)$ are known, so we do not need to consider them here). We let $\beta = 5{,}400' - \alpha$, so that β is the complement to the angle α. Since

$$\mathrm{Sin}(\beta) = \mathrm{Sin}(90° - \alpha) = \mathrm{Cos}(\alpha), \tag{87}$$

we can find the Cosine of α by computing the Sine of β. The following procedure is meant to be carried out for both α and β, but we will present it only in the case of α.

Let q and r be the quotient and remainder, respectively, of the division of α by 225. Then $\alpha = 225 \cdot q + r$ where $0 \leq r < 225$. Furthermore, we have that $\mathrm{Sin}_q \leq \mathrm{Sin}(\alpha) < \mathrm{Sin}_{q+1}$. Jñānarāja calls Sin_q the traversed Sine and Sin_{q+1} the present Sine. We can now compute $\mathrm{Sin}(\alpha)$ as

$$\mathrm{Sin}(\alpha) = \mathrm{Sin}_q + \frac{r}{225} \cdot (\mathrm{Sin}_{q+1} - \mathrm{Sin}_q). \tag{88}$$

As noted, this is straightforward linear interpolation.

(12) **[Start by] subtracting the [greatest possible] Sine [given in the table from the given Sine]. Whatever is the result of the remainder [of the subtraction of the two Sines] divided by the difference between the traversed and the current Sines, the angle [corresponding to the given Sine] is [equal to] that increased by 225 multiplied by the number [corresponding to] how large the subtracted Sine was.**

The Sine table can also be used to find the angle corresponding to a given Sine. As in the previous verse, the method used is simple linear interpolation. The procedure is as follows.

Suppose that we are given the value of a Sine, say, $\mathrm{Sin}(\alpha)$, and want to compute the angle α, which is assumed to be between 0° and 90°. We start by finding the greatest positive integer m that satisfies $\mathrm{Sin}_m \leq \mathrm{Sin}(\alpha) < \mathrm{Sin}_{m+1}$. As before, Sin_m is called the traversed Sine and Sin_{m+1} the current Sine. The number corresponding to the Sine that we subtracted, i.e., Sin_m, is m. According to the verse, α can be found as

$$\alpha = 225 \cdot \frac{\text{Sin}(\alpha) - \text{Sin}_m}{\text{Sin}_{m+1} - \text{Sin}_m} + 225 \cdot m. \tag{89}$$

In this way, we can compute the angle corresponding to any given Sine.

(13–14) **The [successive] differences of the small Sines are 25, 24, 23, 21, 19, 16, 13, 10, 6, and 3.**

[The procedure for computing the small Sine of a given angle is as follows.] The degrees in the angle are divided by 9. [The quotient of the division] is [the number of] the traversed [small] Sine. The remainder [of the division] is multiplied by the current [small] Sine and divided by 9. [The result is then] increased by the traversed [small Sine] differences. [This is the small Sine of the angle.]

[The procedure for computing the angle corresponding to a given small Sine is as follows.] One should subtract [as many of] the [small Sine] differences [from the given small Sine as possible]. [When] the result from the division of the remainder [of the subtraction process] by the current [small Sine] is increased by 9 multiplied by the number corresponding to the traversed [small Sine], [the result] is the angle.

Jñānarāja here presents another Sine table. It is based on $R = 160$ and a division of the interval between 0° and 90° into 10 equal parts of 9° each. These Sines are called small Sines (*laghujyā*) to distinguish them from the Sines given before in verses 2–5, and we shall use the notation $\text{Sin}^\ell(\alpha)$ to indicate the small Sine of the angle α. We then have that $\text{Sin}^\ell(\alpha) = 160 \cdot \sin(\alpha)$ if $0° \leq \alpha \leq 90°$. As before, we define Sin^ℓ_n as $\text{Sin}^\ell(n \cdot 9°)$ for $0 \leq n \leq 10$ and ΔSin^ℓ_n as $\text{Sin}^\ell_n - \text{Sin}^\ell_{n-1}$ for $1 \leq n \leq 10$.

What is tabulated here by Jñanaraja is not small Sines directly, but rather differences of small Sines. As such, the numbers

Sine number	Angle	ΔSin_n^ℓ	Sin_n^ℓ	Value	Difference
1	9°	25	25	25.029	25.029
2	18°	24	49	49.442	24.413
3	27°	23	72	72.638	23.195
4	36°	21	93	94.045	21.407
5	45°	19	112	113.137	19.091
6	54°	16	128	129.442	16.305
7	63°	13	141	142.561	13.118
8	72°	10	151	152.169	9.607
9	81°	6	157	158.030	5.861
10	90°	3	160	160.000	1.969

Table 7: Table of small Sine differences

tabulated are, in order, $\Delta\text{Sin}_1^\ell, \Delta\text{Sin}_2^\ell, \ldots, \Delta\text{Sin}_{10}^\ell$. Note that

$$\text{Sin}_n^\ell = \sum_{k=1}^{n} \Delta\text{Sin}_k^\ell, \tag{90}$$

a result that is easily verified.

The reason for including the table of small Sine differences is that these are easier to do computations with than the Sines. The numbers are smaller and there are fewer of them. The values produced by small Sine differences are less accurate than those produced by the Sines, but not significantly so for practical purposes. In the verses containing problems and their solutions in the next section, Jñānarāja uses small Sines rather than Sines for his computations.

Table 7 gives for each successive small Sine difference its number, its corresponding angle, its value, its corresponding small Sine (not given in the text), a modern computation of the small Sine, and a modern computation of the small Sine difference.

As with the previous Sine table, this table can be used to compute the small Sine of a given angle, as well as the angle corresponding to a given small Sine. Yet again, the method used is linear interpolation.

Let there be given an angle, α, and let q and r be the quotient and the remainder, respectively, of the division of α by 9. Then $\alpha = 9 \cdot q + r$, where $0 \leq r < 9$. Furthermore, q is the number of the traversed small Sine, Sin^ℓ_q, as mentioned in the verses. The small Sine of α can now be found as

$$\mathrm{Sin}^\ell(\alpha) = \frac{r}{9} \cdot \Delta\mathrm{Sin}^\ell_{q+1} + \sum_{k=1}^{q} \Delta\mathrm{Sin}^\ell_k. \tag{91}$$

Now, if instead of an angle we are given the value of a small Sine, say, $\mathrm{Sin}^\ell(\alpha)$, we can find the angle as follows. Let n be the largest possible integer that satisfies that $r = \mathrm{Sin}^\ell(\alpha) - \sum_{k=1}^{n} \Delta\mathrm{Sin}^\ell_k$ is nonnegative. Put differently, we subtract as many small Sine differences as possible from $\mathrm{Sin}^\ell(\alpha)$. The result is r. Then α can be found as

$$\alpha = \frac{9 \cdot r}{\Delta\mathrm{Sin}^\ell_{n+1}} + 9 \cdot n. \tag{92}$$

Yet again this is simple linear interpolation.

(15) **The *manda* anomaly is [defined as the longitude of] the [mean] planet diminished by [the longitude of] the *manda* apogee, [and] the *śīghra* anomaly is [defined as the longitude of] the *śīghra* apogee diminished by [the longitude of] the [mean] planet.**

If [either] the *śīghra* or the *manda* anomaly is situated in [the half circle] beginning with Aries or in [the half circle] beginning with Libra, the respective equations are a positive and a negative application or a negative and a positive application, respectively.

If the *śīghra* anomaly is situated in [the half circle] beginning with Aries or in [the half circle] beginning with Libra, its equation is a positive or a negative application [respectively]. [Similarly,] if the *manda* anomaly [is situated in the half circle beginning with Aries or in the half circle beginning with Libra], [its equation] is a negative or positive application [respectively].

Having concluded the section on Sines, Jñānarāja turns to planetary theory. At this point, it will be instructive to give a brief description of the Indian planetary model.

The planetary model of the Indian tradition is illustrated in Figure 10.[192] The center of each planetary orbit is the center of

[192]See also Pingree 1978a, 557–558.

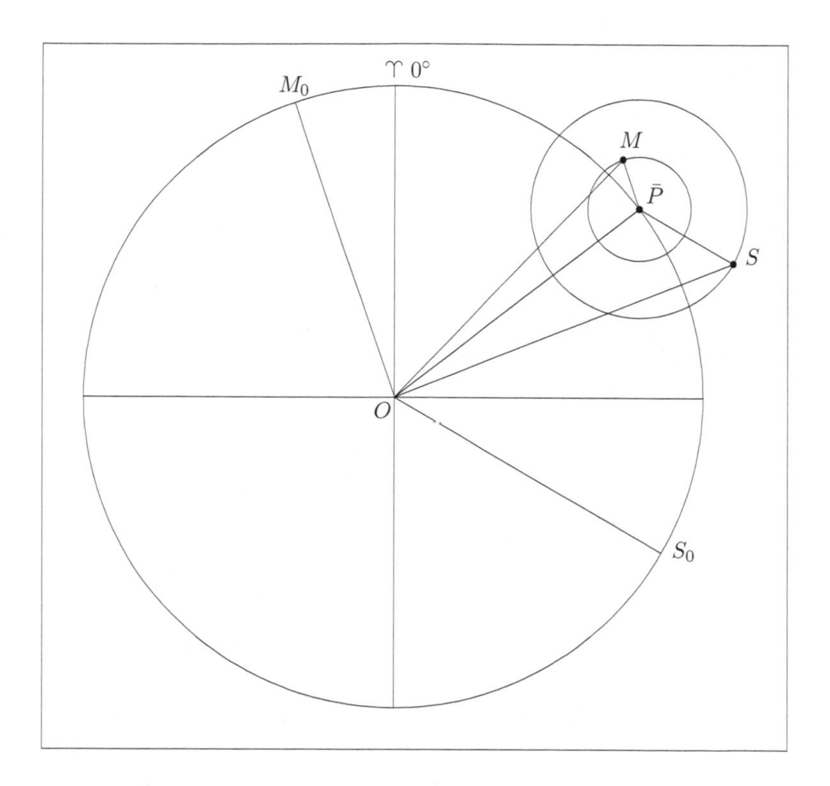

Figure 10: The epicyclical model of Indian astronomy

the earth, which is marked by O in the figure. The mean planet, $\bar{P}$, which was described in the previous section, rotates around O on the large circle of the figure, called the deferent. The deferent is divided into four quadrants consisting of three signs each, and planetary longitudes are measured from the beginning of Aries, the first of the 12 signs. The beginning point of Aries is marked as $\Upsilon\ 0°$ in the figure.

Centered around $\bar{P}$ are, in the case of the five star-planets, two epicycles: the *manda* (slow) epicycle, which is the smaller of the two in the figure, and the *śīghra* (fast) epicycle, which is the larger one. For the sun and the moon, there is only one epicycle: the *manda* epicycle. The circumferences of the two vary for each star-planet, but with the exception of Saturn, the *śīghra* epicycle is always larger than the *manda* epicycle.

Associated with each epicycle is an apogee: M_0 in the case of the *manda* epicycle, and S_0 in the case of the *śīghra* epicycle. The point M on the *manda* epicycle and the point S on the *śīghra* epicycle satisfy that the lines $\bar{P}M$ and $\bar{P}S$ are parallel to OM_0 and OS_0, respectively. The points where the lines OM and OS, extended if necessary, intersect the deferent would be the position of the planet if the respective epicycle acted alone. For the sun and the moon, the point of intersection between the line OM and the deferent is the position of the true planet. In the case of the five star-planets, the true planet is found by combining the two epicycles in a manner that will be explained later in the section (see verses $23^{\text{c–d}}$–24).

The *manda* equation of center, or simply equation, is defined as the angle $\bar{P}OM$, and the *śīghra* equation is defined as the angle $\bar{P}OS$. The equation is thus the angle by which the mean planet is displaced by the respective epicycle.

The *manda* anomaly, κ_μ, is defined as the difference between the longitude of the mean planet and the longitude of the *manda* apogee, and the *śīghra* anomaly, κ_σ, is defined as the difference between the longitude of the *śīghra* apogee and the longitude of the mean planet. In both cases, the anomaly is the angle between the respective apogee and the mean planet, and in the figure they are $\angle M_0O\bar{P}$ and $\angle S_0O\bar{P}$, respectively; the difference in the definitions only affects when the equation is to be applied positively or negatively. That the *manda* equation is positive between 0° and 180° and negative between 180° and 360°, while

Planet	*Manda* anomaly		*Śīghra* anomaly	
	0°, 180°	90°, 270°	0°, 180°	90°, 270°
The sun	14°	13;40°	—	—
The moon	32°	31;40°	—	—
Mars	75°	72°	235°	232°
Mercury	30°	28°	133°	132°
Jupiter	33°	32°	70°	72°
Venus	12°	11°	262°	260°
Saturn	49°	48°	39°	40°

Table 8: Epicyclical circumferences at the end of quadrants

it is opposite for the *śīghra* equation, is clear from the figure and the definitions of the two anomalies.

(16–18[a–b]) **When the [*manda*] anomaly is at the end of an even quadrant, the degrees of the circumferences in the *manda* epicycle are 14 for the sun, 32 for the moon, 75 in the case of Mars, 30 in the case of Mercury, 33 for Jupiter, 12 for Venus, [and] 49 for Saturn.**

When the [*manda*] anomaly is at [the end of] an odd quadrant, [the degrees of the circumferences in the *manda* epicycle are] 13;40 for the sun, 31;40 for the moon, and 72, 28, 32, 11, and 48 for [the star-planets] beginning with Mars.

[When the *śīghra* anomaly is] at [the end of] an even quadrant, the degrees in the *śīghra* epicycle are 235, 133, 70, 262, and 39 for [the star-planets] beginning with Mars.

[When the *śīghra* anomaly is] at [the end of] an odd quadrant, [the degrees in the *śīghra* epicycle] are 232, 132, 72, 260, and 40 for [the star-planets] beginning with Mars.

The two even quadrants are the second and the fourth, and the two odd quadrants are the first and the third. When the anomaly is at the end of an even quadrant, it is therefore either 0° or 180°. It is similarly either 90° or 270° at the end of an odd quadrant. The circumferences of the epicycles when the anomalies have these values are given in Table 8. Note that the circumferences are measured in units of which there are 360 in the deferent, which is why their their values are given in degrees.

($18^{\text{c–d}}$) **[After] the Sine of the anomaly is multiplied by the difference between the circumferences [at the end of even and odd quadrants, respectively,] and divided by the radius, the circumferences [at the end of even quadrants] are diminished or increased by the result according to whether they are greater or less than [the circumferences at the end of] odd [qudrants]. [These are the] true [circumferences].**

Let c_2 be the circumference of either the *manda* epicycle or the *śīghra* epicycle of a given planet at the end of an even quadrant, and let c_1 be the circumference at the end of an odd quadrant. Let d be the difference between c_2 and c_1, and remember that in the Indian system such a difference is always a positive number. In other words, in our notation we have that $d = |c_2 - c_1|$. Assume further that the current anomaly is κ. Finally, let b be given by $b = \frac{\text{Sin}(\kappa) \cdot d}{R}$. Note that b is just a linear interpolation factor.

The epicyclical circumference, c, corresponding to the anomaly κ is given by

$$c = \begin{cases} c_2 - b & \text{if } c_2 > c_1, \text{ and} \\ c_2 + b & \text{if } c_2 < c_1. \end{cases} \tag{93}$$

This formula is given in the *Sūryasiddhānta*.[193]

(19) **[When] the Cosine and the Sine [of the anomaly] are multiplied by the true circumference and divided by 360, [the two results] are the *koṭiphala* and the *bhuja-phala*.**

The arc corresponding to the *manda bhujaphala* is [approximately] the *manda* equation of the planet in minutes of arc.

[193] *Sūryasiddhānta* 2.38. It is also given by other astronomers; for example by Brahmagupta in *Brāhmasphuṭasiddhānta* 2.13.

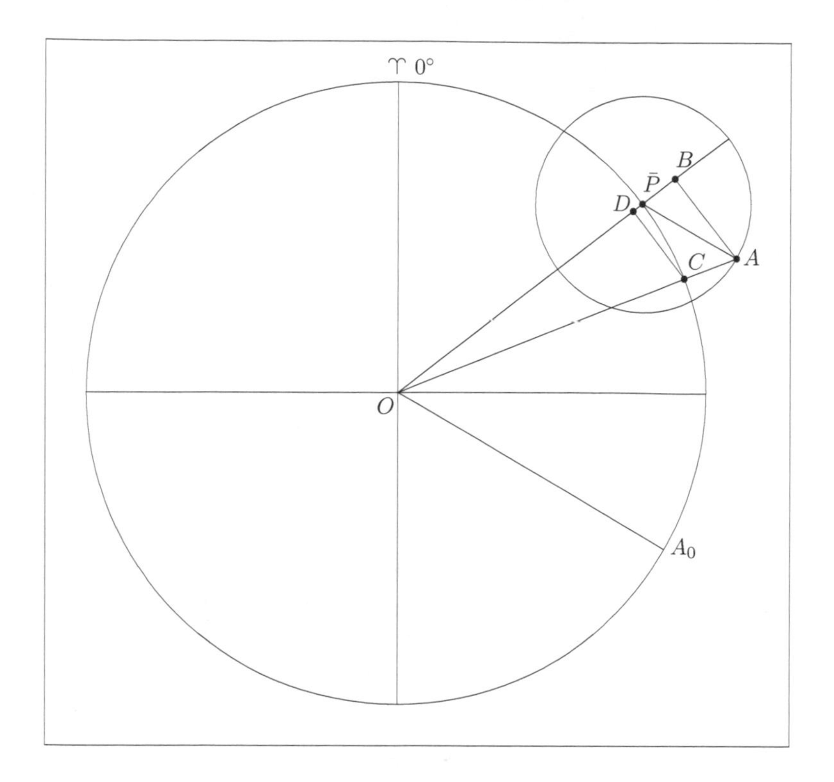

Figure 11: The *koṭiphala* and the *bhujaphala*

As noted, the circumference, c, of an epicycle as given in verses 16–18$^{\text{a–b}}$ is measured in units of which there are 360 in the deferent circle.[194] Since the circumference of the deferent is $2 \cdot \pi \cdot R$, the circumference of a given epicycle, measured in regular units, is $\frac{2 \cdot \pi \cdot R}{360} \cdot c$. Its radius, r, is then

$$r = \frac{1}{2 \cdot \pi} \cdot \frac{2 \cdot \pi \cdot R}{360} \cdot c = \frac{c}{360} \cdot R. \tag{94}$$

This shows that the ratio $\frac{c}{360}$ simply converts lengths related to the deferent to the corresponding lengths related to the epicycle.

Now, Figure 11 depicts the deferent and an epicycle (let us assume that it is a *manda* epicycle) with apogee A_0. Let κ be the anomaly, i.e., $\angle A_0 O \bar{P}$. The line AB is drawn so that it is perpendicular to the line OP_0.

The *bhujaphala* and the *koṭiphala* are, respectively, the line segments AB and BP_0. It is clear from the figure and from the

[194]See Pingree 1978a, 558.

fact that the factor $\frac{c}{360}$ converts lengths related to the deferent to the corresponding lengths related to the epicycle that the length of the *koṭiphala* is $\frac{c}{360} \cdot \text{Cos}(\kappa)$ and that of the *bhujaphala* is $\frac{c}{360} \cdot \text{Sin}(\kappa)$.

As is customary in the Indian astronomical tradition, Jñānarāja takes the Sine of the *manda* equation to be the length of the *bhujaphala*. As can be easily seen from the figure, this is not correct, for the Sine of the *manda* equation is the length of CD, whereas the *bhujaphala* is the segment AB. However, as the circumferences of the *manda* epicycles are small, it is an acceptable approximation. The circumferences of the *śīghra* epicycles are generally too large for the approximation to be useful.

Note that Jñānarāja's use of Sine and Cosine implies that the deferent circle is treated as the standard trigonometric circle. In other words, it is assumed to have a radius of 3,438.

(20) **Alternatively, the small Sine of the anomaly is multiplied by 4 and divided by 67. The result, in minutes of arc and so on, is multiplied by the [true] circumference [of the epicycle]. [The resulting quantity is the *bhujaphala*.]**

For the sake of ease [the arc corresponding to] this [*bhujaphala*] is called the *manda* [equation].

From the previous verse, we know that the *bhujaphala* is equal to $\frac{c}{360} \cdot \text{Sin}(\kappa)$. If we want to express the *bhujaphala* in terms of a small Sine rather than a Sine (that is, to use the radius $R = 160$ rather than the radius $R = 3{,}438$), we need the number x that satisfies $\frac{3{,}438}{360} = \frac{160}{x}$. For then

$$\frac{c}{360} \cdot \text{Sin}(\kappa) = c \cdot |\sin(\kappa)| \cdot \frac{3{,}438}{360} = c \cdot |\sin(\kappa)| \cdot \frac{160}{x} = c \cdot \text{Sin}^{\ell}(\kappa) \cdot \frac{1}{x}, \tag{95}$$

and we can express the *bhujaphala* in terms of a small Sine rather than a Sine.

By expanding $\frac{1}{x}$ as a continued fraction, we get

$$\frac{1}{x} = \frac{3{,}438}{360 \cdot 160} = \frac{191}{3{,}200} = \frac{1}{16 + \frac{1}{1 + \frac{1}{3 + \frac{3}{7}}}} \approx \frac{1}{16 + \frac{1}{1 + \frac{1}{3}}} = \frac{4}{67}, \tag{96}$$

from which the formula follows.

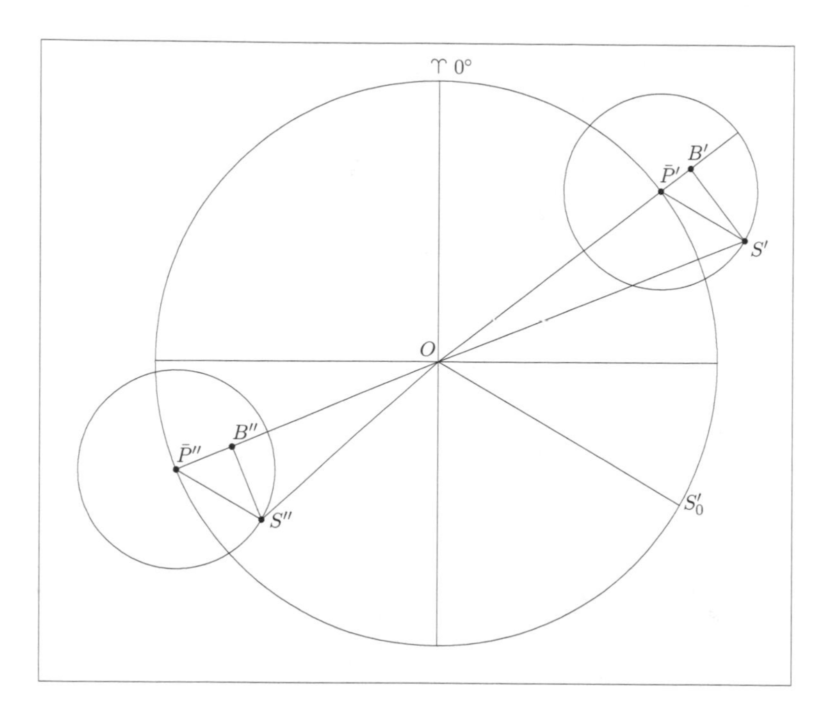

Figure 12: The hypotenuse

(21) **The sum or the difference of the radius and the *śīghra koṭiphala*, when the anomaly is in [one of the six signs] beginning with Capricorn or in [one of the six signs] beginning with Cancer, respectively, is to be computed. The square root of the sum of the square of [the result of] that [operation] and the square of the *bhujaphala* is the hypotenuse, measured by the distance between the center of the earth and the planet.**

Figure 12 shows two different positions of the *śīghra* epicycle. For each position, a right-angled triangle is formed, namely, the triangle $OS'B'$ and the triangle $OS''B''$. In the first case, the radius of the deferent (which is $R = 3{,}438$) is increased by the *koṭiphala*, and in the second case, it is decreased by the *koṭiphala*. According to the verse, the *koṭiphala* is added when the anomaly is either between 0° and 90° or between 270° and 360°, and subtracted when the anomaly is between 90° and 270°. This can be verified easily from the figure.

The hypotenuse (*karṇa*) represents the distance between the center of the earth and the position the planet would have if the *śīghra* epicycle acted alone on it. Since the *śīghra* epicycle has a much greater effect than the *manda* epicycle, the hypotenuse is roughly the geocentric distance of the planet. If we denote the hypotenuse by H, the *koṭiphala* by k, and the *bhujaphala* by b, then

$$H = \sqrt{(R \pm k)^2 + b^2} \tag{97}$$

by the Pythagorean theorem. On the figure, depending on which epicycle we are dealing with, H is the length of OS' or OS'', k is the length of $B'\bar{P}'$ or $B''\bar{P}''$, and b is the length of $B'S''$ or $B''S''$.

(22) **An [approximate value of the desired] square root is increased by the result of the division of the given square [i.e., the number whose square root we are seeking] by the approximate square root. [The result] is divided by 2. That is the [new] approximate square root. Then [the process is repeated] again and again. In this way, the correct square root [is found].**

The previous verse required the computation of a square root, and so a procedure to carry out such a computation is given in the present verse.

The method works as follows. Suppose that we want to compute the square root of the positive number A. Suppose further that the positive number a_0 is an approximation to $\sqrt{A}$, i.e., that $A - a_0^2$ is small. If we let

$$a_1 = \frac{1}{2} \cdot \left(a_0 + \frac{A}{a_0}\right), \tag{98}$$

then a_1 is a better approximation to $\sqrt{A}$. Continuing this process iteratively by letting

$$a_{n+1} = \frac{1}{2} \cdot \left(a_n + \frac{A}{a_n}\right) \tag{99}$$

for $n = 1, 2, 3, \ldots$, we obtain a sequence,

$$a_0, a_1, a_2, \ldots, a_n, a_{n+1}, \ldots,$$

of successively better and better approximations to $\sqrt{A}$. When sufficient accuracy has been achieved, that is, when a_n and a_{n+1} are sufficiently close to each other, the process is stopped and the square root is then determined as $\sqrt{A} = a_{n+1}$.[195]

This verse is also found in the *bījagaṇitādhyāya*,[196] from which it has been quoted in the secondary literature.[197] It is quoted (from the *bījagaṇitādhyāya*) by Sūryadāsa in the *Sūrya-prakāśa*.[198]

Historical notes regarding this method were given in the Introduction (see page 31).

(23^{a-b}) **The *bhujaphala* is multiplied by the radius and divided by the *śīghra* hypotenuse. The minutes in the arc corresponding to the result of that [operation] provide the *śīghra* equation.**

In Figure 13 the triangle OBS is similar to the triangle ODC. If σ denotes the *śīghra* equation, i.e., $\angle DOC$, then clearly $\text{Sin}(\sigma) = |CD|$. In addition, since the triangles OBS and ODC are similar, we have that

$$\text{Sin}(\sigma) = \frac{b \cdot R}{H}, \tag{100}$$

where H is the length of OS and b is the length of BS, that is, they are the hypotenuse and the *bhujaphala*, respectively.

(23^{c-d}–24) **The sun and the moon are corrected by [one] entire [application of the] *manda* [equation]. The others are corrected by four [applications of] equations.**

Having applied half of the [planet's] own *śīghra* equation positively or negatively [as the case may be] to the mean [planet], then half the *manda* equation [found from the new position is applied] to it [i.e., to the new

[195] That the sequence always converges, and fast, to $\sqrt{A}$ is well known from modern mathematics. The method is, in fact, equivalent to applying the Newton-Raphson method to the function f defined by $f(x) = x^2 - A$ for all $x > 0$.

[196] See Jain 2001, 12–13 of Introduction.

[197] See, e.g., Bag 1979, 101.

[198] See Jain 2001, 45 of Sanskrit text.

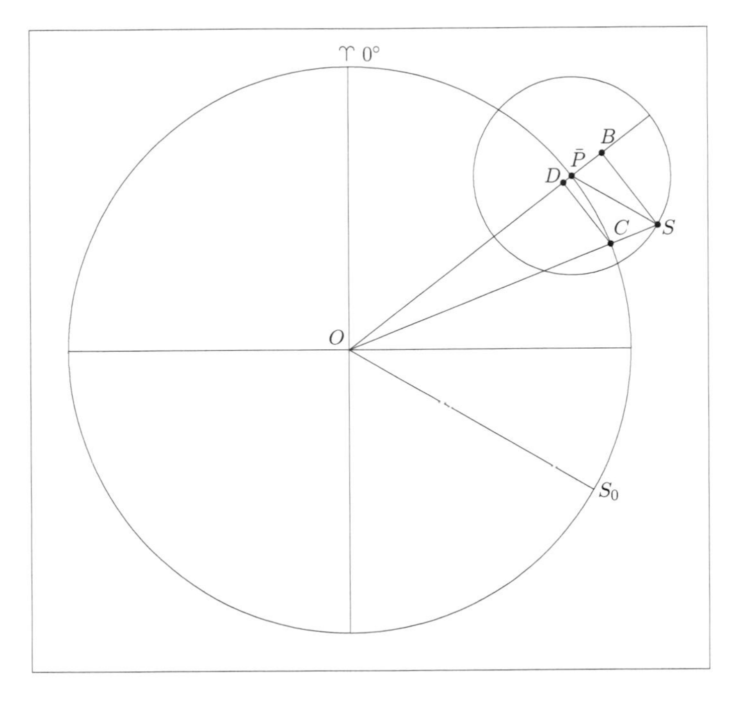

Figure 13: The *śīghra* equation

position]. [This gives yet another position.] By [an application to the mean planet of] the entire *manda* [equation] derived from that [position], the mean [planet] becomes *manda*-corrected. [The *manda*-corrected planet] becomes [the] true [planet] by [an application of] the entire *śīghra* [equation].

The sun and the moon have only one epicycle each, namely, the *manda* epicycle. To get their true positions, one need only apply (that is, add or subtract, as the situation may be) the *manda* equation to the mean position. This is clear and simple.

The five star-planets, on the other hand, have two epicycles, the *manda* epicycle and the *śīghra* epicycle. From Figure 10 it may appear as if the two epicycles are independent, but in reality they are not. They need to be combined, but how to do so is not obvious. The procedure followed by Jñānarāja consists of four applications of equations.

First, half of the *manda* equation is applied to the mean planet. This gives a new position. Next, half of the *śīghra* equation computed from the new position is applied to the new position. Then the *manda* equation computed from that position is applied to that position, at which point the planet is called *manda*-corrected. Finally, the entire *śīghra* equation is applied to the *manda*-corrected planet, which gives the true position of the planet. This is the same procedure as given in the *Sūryasiddhānta*.[199]

For a recent discussion of the rationale behind this procedure, see Duke 2005.

(25) **The *śīghra* equation computed from [the mean planet corrected by] the entire *manda* equation is not correct. First, the *manda*-corrected [planet] is to be known from the mean planet only. Then, the *manda* equation produced from the mean [planet] corrected by half of the *śīghra* equation and [half of] the *manda* equation is correct. [That is] sufficient. What is strange in this procedure?**

The verse restates the procedure previously given, adding

[199] *Sūryasiddhānta* 2.43–44.

that it is incorrect to merely apply the whole *manda* equation and then the whole *śīghra* equation to get the true planet.

(26–28) **The velocity of the *manda* anomaly is [first] multiplied by the difference of the Sines [i.e., the difference between the current Sine and the previous Sine], [next] divided by the first half-Chord, [then] multiplied by the true circumference, [and finally] divided by the degrees in a rotation [i.e., by 360]. [This is] the result for the *manda* velocity.**

The [mean] velocity [of a planet] diminished or increased by that result, according to whether the [*manda*] anomaly is in [one of the six signs] beginning with Capricorn or in [one of the six signs] beginning with Cancer, is *manda*-corrected.

Having subtracted that [*manda*-corrected velocity] from the velocity of the *śīghra* apogee, the remainder is the motion produced by the *śīghra* anomaly.

In the case of the [effect of the] course of the arc of the *śīghra* equation [on the velocity], this [remainder] is multiplied by the difference of the Sines [i.e., the difference between the current Sine and the previous Sine] and divided by the first Sine. [Then the result is] divided by the [*śīghra*] hypotenuse and multiplied by the radius. [The resulting quantity] is subtracted from the velocity of the *śīghra* apogee, [yielding] the true velocity [of the planet].

Let κ_μ and κ_σ be the *manda* anomaly and the *śīghra* anomaly, repectively. Let v_κ and w_κ be the velocities of the *manda* anomaly and the *śīghra* anomaly (i.e., the rate by which each changes during a civil day), respectively. Assume further that $n \cdot 3°45' \leq \kappa_\mu < (n+1) \cdot 3°45'$. Then Sin_{n+1} is the current Sine and Sin_n is the previous Sine in the case of the *manda* anomaly.

The mean planet moves at a constant velocity, but due to the effects of the epicycles (or epicycle, in the case of the luminaries), the velocity of the true planet, i.e., the true velocity, changes over time. Both the *manda* epicycle and the *śīghra* epicycle contribute to this.

In order to find the effect of the *manda* epicycle on the velocity, the following quantity is computed:

$$u = v_\kappa \cdot \frac{\mathrm{Sin}_{n+1} - \mathrm{Sin}_n}{\mathrm{Sin}_1} \cdot \frac{c}{360}. \tag{101}$$

Why u has this form is not clear.[200] This is to be applied to the mean velocity to get the velocity after the effect of the *manda* epicycle has been applied.

If the *manda* anomaly is in one of the six signs beginning with Capricorn, i.e., if it is between 270° and 360° or between 0° and 90°, the displacement of the mean planet by the *manda* epicycle has the effect that the true velocity is smaller than the mean velocity. Therefore, the *manda*-corrected velocity, v_m, in this case is

$$v_m = \overline{v} - u. \tag{102}$$

If the *manda* anomaly is in one of the six signs beginning with Cancer, i.e., if it is between 90° and 270°, it is opposite, so that

$$v_m = \overline{v} + u. \tag{103}$$

Assume that $m \cdot 3°45' \leq \kappa_\sigma < (m+1) \cdot 3°45'$. Then Sin_{m+1} is the current Sine and Sin_m is the previous Sine in the case of the *śīghra* anomaly.

Let w_σ be the velocity of the *śīghra* apogee and H the hypotenuse (i.e., the geocentric distance of the planet). The correction to the velocity due to the *śīghra* epicycle is computed as follows:

$$u' = (w_\sigma - v_m) \cdot \frac{\mathrm{Sin}_{m+1} - \mathrm{Sin}_m}{\mathrm{Sin}_1} \cdot \frac{R}{H}. \tag{104}$$

The form of u' is also not clear.[201] If this quantity is smaller than the velocity of the *śīghra* apogee, then the true velocity of the planet, v, is

$$v = w_\sigma - u'. \tag{105}$$

However, if it is larger, then the planet is in retrograde motion with the velocity

$$v = u' - w_\sigma. \tag{106}$$

This will be taken up in the following by Jñānarāja.

---

[200] See Pingree 1978a, 569.

[201] See Pingree 1978a, 569.

(29–31) **When that [correction due to the *śīghra* epicycle] cannot be subtracted [from the *manda*-corrected velocity due to the latter being smaller than the former] it is subtracted inversely [i.e., the former is subtracted from the latter], and the remainder [of that subtraction] is the retrograde velocity of the [star-]planets.**

It has been said by the great ancients that [since] the sun and the moon have no *śīghra* epicycles, therefore [they] never [have] retrogradation.

[When] a [star-]planet that is located very far from its *śīghra* apogee commences [its] retrograde motion, then, starting with Mars, these are the degrees of the *śīghra* anomaly: 164, 144, 130, 163, **and** 115.

The degrees at which [the star-planets commence their] direct motion are 360 **[degrees] diminished by those.**

The degrees and minutes of arc greater or less than [these] given values [for the occurrence of the first station] divided by the velocity of the [*śīghra*] anomaly are the days that have elapsed since or are to pass until [the previous or the next station].

For retrogradation and the stations of the star-planets, see the commentary on 2.1.39.

When the correction to the velocity due to the *śīghra* epicycle cannot be subtracted from the *manda*-corrected motion, the direction of the motion is opposite the usual direction, i.e., the planet is in retrograde motion. In this case, one subtracts the *manda*-corrected velocity from the *śīghra* correction and notes that the planet is retrograde.

The occurrences of the stations are determined by the *śīghra* anomalies. Table 9 gives the values of the *śīghra* anomalies at which each star-planet reaches its first and second stations. These values are the same as those given in the *Sūryasiddhānta*.[202]

If we want to compute the time elapsed since the last occurrence of the first station or the time to pass before the next occurrence of the first station, Jñānarāja tells us to take the angular distance between the current location and the past or future occurrence of the first station and divide by the velocity of the *śīghra* anomaly. This is a straightforward formula.

[202] *Sūryasiddhānta* 2.53–54. See also Pingree 1978a, 610.

Planet	First station	Second station
Mars	164°	196°
Mercury	144°	216°
Jupiter	130°	230°
Venus	163°	297°
Saturn	115°	245°

Table 9: *Śīghra* anomalies at the first and second stations

(32^{a-b}) **The *manda* equation of the sun [*bāhuphala*] is multiplied by the velocity [of a given planet] and divided by the minutes of arc in all the *nakṣatras* [i.e., by 21,600]. [The result] is applied positively or negatively to the [longitude of the given] planet according to [whether] the equation [of the sun [*doḥphala*] is to be applied positively or negatively]. From this is [found the longitude of the given planet] at the rising of the true sun.**

The words *bāhuphala* and *doḥphala* must here refer to the *manda* equation of the sun. If so, the formula is given for the purpose of computing what is known as the *bhujāntara* in Indian astronomy.[203]

In treatises following a sunrise system, the computed planetary positions correspond to the rising of the mean sun. In order to find their longitudes when the true sun is rising, the *bhujāntara* correction is applied. However, the inclusion of the formula in this form is somewhat curious, as Jñānarāja follows a midnight system, not a sunrise system.

The *nakṣatras* are constellations along the path of the moon. Together they span 360° and therefore contain $60 \cdot 360 = 21{,}600$ minutes of arc.

(32^{c-d}) **The Sine produced from the precessional [longitude of the] sun is multiplied by the Sine of 24 [degrees] and divided by the radius. [The result] is the Sine of the declination [of the sun]. The arc corresponding to this**

[203] See, e.g., *Brāhmasphuṭasiddhānta* 2.29 and *Śiṣyadhīvṛddhidatantra* 2.16. See also Pingree 1978a, 569.

[Sine] is the declination [of the sun] having the direction of the hemisphere [that the sun is in].

Let δ and λ^* be the declination of the sun and the precessional longitude of the sun, respectively. For the formula given here,

$$\mathrm{Sin}(\delta) = \frac{\mathrm{Sin}(\lambda^*) \cdot \mathrm{Sin}(\varepsilon)}{\mathrm{Sin}(90^\circ)} = \frac{\mathrm{Sin}(\lambda^*) \cdot \mathrm{Sin}(\varepsilon)}{R}, \tag{107}$$

see the commentary on $2.1.29^{\text{b–d}}$. If the sun is north of the celestial equator, the direction of the declination is north, and similarly if the sun is south of the celestial equator, the direction of the declination is south.

Note that precessional longitude of the sun (or another planet) is simply the tropical longitude, i.e., the longitude with reference to the vernal equinox.

(33) **[First] the Sine of the precessional [longitude of the] sun is increased by [its own] 61st part, and [then the result is] divided by 10. Good people call this [quantity] multiplied by 4 the Sine of the declination.**

The word *anupamaguṇa* in the verse is unclear, and it is not included in the above translation. Perhaps it means that the Sine of the declination is matchless, though this makes little sense. It is also possible that it means that this simplified formula gives a perfectly good approximation. At any rate, the formula given is clear.

This formula is a simplified one for computing the declination of the sun, namely,

$$\mathrm{Sin}(\delta) = 4 \cdot \frac{\mathrm{Sin}(\lambda^*) + \frac{1}{61} \cdot \mathrm{Sin}(\lambda^*)}{10} = \frac{124}{305} \cdot \mathrm{Sin}(\lambda^*), \tag{108}$$

where δ is the declination of the sun and λ^* is the tropical longitude of the sun. It is derived from the previous one by the following approximation:

$$\frac{\mathrm{Sin}(\varepsilon)}{R} = \frac{1{,}397}{3{,}438} = \frac{124}{305} - \frac{227}{1{,}048{,}590} \approx \frac{124}{305}. \tag{109}$$

(34) **The square root of the difference between the square of the radius and the square of the Sine of the declination [of the sun] is the radius of the [sun's] diurnal circle. The Sine of the declination [of the sun] multiplied by the noon equinoctial shadow and divided by 12 is the earth Sine. The earth Sine divided by the radius of the [sun's] diurnal circle and multiplied by the radius is the Sine of the ascensional difference. The arc of that [Sine of the ascensional difference] is the rising and setting ascensional difference. [The existence] of these two [i.e., the rising and setting ascensional differences] is the difference between [a region] without latitude and [a region] with latitude [for there is no ascensional difference in the former].**

The formulae given here and the similar triangles that they are based on will be discussed in greater detail in the next section. Definitions and further details will be given there (see verses 2.3.6–11 and commentary).

Let r be the radius of the sun's diurnal circle and δ the declination of the sun. Then

$$r = \sqrt{R^2 - (\mathrm{Sin}(\delta))^2}. \tag{110}$$

If s_0 is the equinoctial shadow at noon in a given location and e the earth Sine, then

$$e = \mathrm{Sin}(\delta) \cdot \frac{s_0}{12}. \tag{111}$$

Let ω be the ascensional difference. Then

$$\mathrm{Sin}(\omega) = e \cdot \frac{R}{r}. \tag{112}$$

Clearly, the arc corresponding to $\mathrm{Sin}(\omega)$ is the ascensional difference. The existence of the ascensional difference distinguishes regions not on the terrestrial equator from those on it, which do not have ascensional difference at any time.

($35^{\text{a–b}}$) **The true velocity of a planet is multiplied by the *palas* in the ascensional difference and divided by 60. The seconds of arc attained [as the result] are [applied] positively or negatively to [the true longitude of]**

the planet depending on whether it is at the rising or the setting of the sun, as well as on which of the two hemispheres we are in.

The ascensional difference will be denoted by ω in the following. As it lies on the equator, it is measured in units of time.

Let v be the true velocity of a given planet and ω the ascensional difference. In order to compute a planet's true longitude at local sunrise or sunset, Jñānarāja tells us to compute the quantity

$$\frac{v \cdot \omega}{60}. \tag{113}$$

The true velocity of the planet v is measured in minutes of arc per civil day, but when divided by 60, it becomes minutes of arc per *ghaṭikā*. Multiplied by an ascensional difference ω measured in *pala*s, the result is measured in seconds of arc. This is the angular distance traveled by the planet during the period of time of the ascensional difference.

Depending on whether we are at sunrise or sunset, as well as what hemisphere we are in, the result is to be added or subtracted from the longitude of the true planet. Although not mentioned here, we also need to take into account whether the sun is north or south of the celestial equator.

If the true longitude of the planet that we started with corresponds to the planet's position when the sun rises or sets on the equator, the result will give us the true longitude of the planet at sunset or sunrise at our given location.

(35^{c-d}) **In the Northern Hemisphere, half of the day and half of the night are said by the knowers of the sphere to be [respectively] 15 *ghaṭikā*s increased or decreased by the ascensional difference [depending on whether the sun is north or south of the celestial equator]. It is opposite in the Southern [Hemisphere].**

On the terrestrial equator, day and night always have equal durations. In other regions, there will be a difference depending on the ascensional difference. When the sun is above the celestial equator, the days will be longer than the nights in the Northern Hemisphere. The length of half a day in this case is exactly 15 *ghaṭikā*s (the constant length of half a day on the celestial equator) increased by the ascensional difference, and the length of a night is 15 *ghaṭikā*s diminished by the ascensional difference. It is opposite when the sun is below the celestial equator. It is further clear that the whole scheme is inverted when we are in the Southern Hemisphere.

(36) **The degrees of [the longitude of] the moon diminished by [the longitude of] the sun are divided separately by 12 and 6. [The two results are, respectively,] the elapsed *tithi*s and the elapsed *karaṇa*s. The *ghaṭikā*s [since the last *tithi* or *karaṇa* or until the next *tithi* or *karaṇa* are found] from the remainder as well as the respective divisor by means of a proportion.**

A *tithi* is the time during which the moon gains 12° over the sun; a *karaṇa* is similarly the time during which the moon gains 6° over the sun, i.e., half a *tithi*.[204] As such there are 30 *tithi*s and 60 *karaṇa*s in a lunar month. The formula given for computing the elapsed *tithi*s or *karaṇa*s is straightforward, as is the outline given for computing the time since the last *tithi* or *karaṇa* or until the next *tithi* or *karaṇa*.

(37) **The [number of the current] *karaṇa* is diminished by 1. The [number that is the] remainder [from this number divided] by 7 [tells us which movable *karaṇa* is the current one]. [If the current *karaṇa* is not movable,] the four fixed [*karaṇa*s], Śakuni, Catuṣpada, Nāga, and so on, [which begins] from the second half of the 14th *tithi* of the dark [*pakṣa*], are to be added.**

The 60 *karaṇa*s of a lunar month are numbered starting with 1, and they are further given names. Table 10 shows the relationship between the numbers and names of the *karaṇa*s.[205] Bava, Bālava, Kaulava, Taitila, Gara, Vaṇij, and Viṣṭi, which occur in that order seven times in a row, are called movable (*adhruva* or *cara*), and Śakuni, Catuṣpada, Nāga, and Kiṃstughna, which are near conjunction, are called fixed (*dhruva* or *sthira*). It is clear from the table that Śakuni begins in the middle of the

[204] For *karaṇa*s, see Pingree 1978a, 546.

[205] The table is based on Pingree 1978a, 546, Table III.18.

Karaṇa number								Name
1								Kiṃstughna
2	9	16	23	30	37	44	51	Bava
3	10	17	24	31	38	45	52	Bālava
4	11	18	25	32	39	46	53	Kaulava
5	12	19	26	33	40	47	54	Taitila
6	13	20	27	34	41	48	55	Gara
7	14	21	28	35	42	49	56	Vaṇij
8	15	22	29	36	43	50	57	Viṣṭi
							58	Śakuni
							59	Catuṣpada
							60	Nāga

Table 10: Numbers and names of the *karaṇas*

14th *tithi* of the dark *pakṣa*, for this *tithi* consists of the 57th and the 58th *karaṇas*.

Suppose that the number of the current *karaṇa* is k and that $1 < k < 58$. Let r be the remainder from the division of $k-1$ by 7. If $r = 1$, the current *karaṇa* is Bava; if $r = 2$, it is Bālava; and so on. In other words, except for $r = 0$, which indicates Viṣṭi, r directly gives the number of the movable *karaṇa* that is current.

If k is one of 1, 58, 59, or 60, the current *karaṇa* is one of the four fixed *karaṇas*. Their order and appearance in a lunar month are described in the verse.

That the verse is elliptical and therefore requires a good deal of knowledge on the part of the reader is evident from the large amount of bracketed text in the translation.

(38) **The degrees of [the longitude of] the moon and the degrees of [the longitude of] the moon increased by [the longitude of] the sun are separately multiplied by 3 and divided by 40. [The two results are, respectively,] the *nakṣatra* and the *yoga*.**

Having subtracted the remainder [of any of the two above quantities] from 800, that [result] is [first] multiplied by 20, [then] multiplied by 60, [and finally] divided by the appropriate velocity [depending on whether we are working with the *nakṣatra* or the *yoga*]. [This is] the

respective *ghaṭikās* [to elapse before the commencement of the next *nakṣatra* or *yoga*].

In addition to being divided into 12 signs, the ecliptic is also divided into 27 *nakṣatras*, which are sometimes translated as "lunar mansions." The first *nakṣatra* starts at Aries 0°, and each spans $\frac{360^\circ}{27} = 13^\circ 20' = 800'$.

The word *nakṣatra* here refers to the period of time it takes the moon to traverse a *nakṣatra*, i.e., to travel 800′. Similarly, a *yoga* is a period of time during which the sum of the distances traveled by the sun and the moon equals 800′.

In the following, let $\lambda_{\odot}$ and $\lambda_{☾}$ be the longitudes of the sun and the moon, respectively.

Since longitudes are measured in minutes of arc per civil day and $\frac{800}{60}$ is the length of a *nakṣatra* in degrees, it is clear that

$$\frac{60}{800} \cdot \lambda_{☾} = \frac{3}{40} \cdot \lambda_{☾} \tag{114}$$

determines the *nakṣatra*, the integer part giving the current *nakṣatra* and the fractional part the position of the moon in that *nakṣatra*.

Completely analogously, the *yoga* is determined as

$$\frac{60}{800} \cdot (\lambda_{\odot} + \lambda_{☾}) = \frac{3}{40} \cdot (\lambda_{\odot} + \lambda_{☾}). \tag{115}$$

The remainder, i.e., the fractional part, of either $\frac{3}{40} \cdot \lambda_{☾}$ or $\frac{3}{40} \cdot (\lambda_{\odot} + \lambda_{☾})$ gives, respectively, the minutes of arc that have been traversed since the start of the current *nakṣatra* or *yoga*; if the remainder is subtracted from 800′, we get the minutes of arc to be traversed before the start of the next *nakṣatra* or *yoga*. If this is multiplied by 60 and divided by the appropriate velocity ($v_{☾}$ in the case of a *nakṣatra*, and $v_{\odot} + v_{☾}$ in the case of a *yoga*), we get the *ghaṭikās* to elapse before the start of the next *nakṣatra* or *yoga* (we can similarly get the *ghaṭikās* elapsed since the start of the current *nakṣatra* or *yoga*, but Jñānarāja does not mention this in the verse). The reason for the multiplication by 20 prescribed in the verse is not clear to me.[206]

[206]The *Vaṭeśvarasiddhānta* (2.6.2) likewise multiplies by 20.

(39–40) **Since the beginning and end points of a sign do not rise at the same place on the horizon on account of the obliqueness of the ecliptic, therefore the rising [times] of the signs are not the same.**

[First] the difference between the square of the Sine of one, two, and three signs [taken separately] and the square of the Sine of the [sun's] declination [is to be computed]. The square root [of each result] multiplied by the radius is divided by the radius of the [sun's] diurnal circle. The arcs corresponding to the results [treated as Sines] are diminished by their respective preceding arc[, if it exists], and thus the rising [times] of [all] the signs, [being these three rising times] placed in regular and inverted order, are attained for a city on the terrestrial equator.

For a given location, the rising times [of the first six signs for a location on the terrestrial equator] are diminished or increased by the ascensional differences corresponding to [the sun's precessional longitude being] one, two, and three signs in the right order [in the case of the first three signs] and put down inversely [in the case of the next three signs]. [The rising times] are inverted for the six [signs beginning] from Libra.

In the above translation, the word *kutas* is interpreted in the sense of *yatas*.

For $k = 1, 2, 3$ compute

$$a_k = \frac{R \cdot \sqrt{(\mathrm{Sin}(k \cdot 30°))^2 - (\mathrm{Sin}(\delta_k))^2}}{\mathrm{Cos}(\delta_k)}, \qquad (116)$$

where δ_1, δ_2, and δ_3 are the declinations of the sun when its precessional longitude is 30°, 60°, and 90°, respectively. $\mathrm{Cos}(\delta_k)$ is the radius of the sun's diurnal circle when the precessional longitude of the sun is δ_k. The correctness of the formula, which is also given by Bhāskara II in the *Siddhāntaśiromaṇi*,[207] can easily be shown using spherical geometry.

After computing a_1, a_2, and a_3, we treat each as a Sine and find the corresponding arcs, α_1, α_2, and α_3. In other words, $a_k = \mathrm{Sin}(\alpha_k)$ for $k = 1, 2, 3$. Note that α_1, α_2, and α_3 are the

[207] *Siddhāntaśiromaṇi*, *gaṇitādhyāya*, *spaṣṭādhikāra*, 54–55.

Sign	Rising time, equator	Rising time, elsewhere
Aries	τ_1	$\tau_1 - \omega_1$
Taurus	τ_2	$\tau_2 - \omega_2$
Gemini	τ_3	$\tau_3 - \omega_3$
Cancer	τ_3	$\tau_3 + \omega_3$
Leo	τ_2	$\tau_2 + \omega_2$
Virgo	τ_1	$\tau_1 + \omega_1$
Libra	τ_1	$\tau_1 + \omega_1$
Scorpius	τ_2	$\tau_2 + \omega_2$
Sagittarius	τ_3	$\tau_3 + \omega_3$
Capricornus	τ_3	$\tau_3 - \omega_3$
Aquarius	τ_2	$\tau_2 - \omega_2$
Pisces	τ_1	$\tau_1 - \omega_1$

Table 11: Rising times of the signs

right ascensions of the three arcs of length 30°, 60°, and 90° measured from the vernal equinox (i.e., the longitudes of the endpoints of these arcs measured with respect to the celestial equator).

Finally, let

$$\tau_1 = \alpha_1, \quad (117)$$

$$\tau_2 = \alpha_2 - \alpha_1, \text{ and} \quad (118)$$

$$\tau_3 = \alpha_3 - \alpha_2. \quad (119)$$

Then $\tau + 1$, τ_2, and τ_3 are the rising times of Aries, Taurus, and Gemini, respectively, for a location on the terrestrial equator. To find the rising times for all the signs for such a location, we have to place these three in regular and inverted order, as prescribed by Jñānarāja. This is most easily explained by looking at Table 11, the second column of which gives the rising time of each sign for a location on the terrestrial equator. It is easy to compute τ_1, τ_2, and τ_3, but Jñānarāja does not give their values.

Now, if we are not on the terrestrial equator, the rising times will be different due to the ascensional difference. Let ω_1, ω_2, and ω_3 be the ascensional differences at the given location corresponding to the sun having the declinations 30°, 60°, and 90°. The rule for taking the ascensional differences into account is illustrated by the third column of Table 11 (the inverted order of the last six signs is reflected when we add and subtract the ascensional differences).

(41) **First, [since] they are located very obliquely by a difference of declinations equal to 11;45 degrees, the risings of Aries and Pisces [occur] in very few *palas*. The risings of Taurus and Aquarius [occur] by a difference of declinations equal to 9 degrees. Being located at a difference of declinations equal to 4 degrees, [the risings] of Gemini and Capricornus [occur] by many [*palas*].**

Let again δ_1 be the declination of the sun when its precessional longitude is 30°, δ_2 the declination of the sun when its precessional longitude is 60°, and δ_3 the declination of the sun when its precessional longitude is 90°.

Then

$$\mathrm{Sin}(\delta_1) = \frac{\mathrm{Sin}(30^\circ)\cdot\mathrm{Sin}(24^\circ)}{R} = \frac{1719\cdot 1{,}397}{3{,}438} = 698\frac{1}{2}, \qquad (120)$$

and by the formula given in verse 12,

$$\delta_1 = \frac{1}{60}\cdot\left(225\cdot\frac{\mathrm{Sin}(\delta_1)-\mathrm{Sin}_3}{\mathrm{Sin}_4-\mathrm{Sin}_3}+3\cdot 225\right) = \frac{6{,}845}{584} \approx 11;45^\circ \qquad (121)$$

(the division by 60 is carried out in order to get a result in degrees). This agrees with what is given in the verse.

Similarly,

$$\delta_2 = \frac{101{,}573{,}597}{5{,}414{,}850} \approx 18;46, \qquad (122)$$

but 18;46 − 11;45 = 7;1, whereas the verse has 9. However, a calculation using a modern calculator gives $\delta_2 = 20;37$, and since $20;37 - 11;45 = 8;52$, this is in better agreement with Jñānarāja's data.

Finally, since δ_3 is the declination of an arc of 90°, it is obvious that $\delta_3 = \varepsilon = 24^\circ$. As $24 - 18;46 = 5;14$, this result also deviates from Jñānarāja's numbers. Using the value of δ_2 arrived at by a modern calculation, however, we get $24 - 20;37 = 3;23$, which is closer to Jñānarāja's data.

It appears that Jñānarāja was able to compute a better value of δ_2 than what the method of verse 12 gives us. His value is,

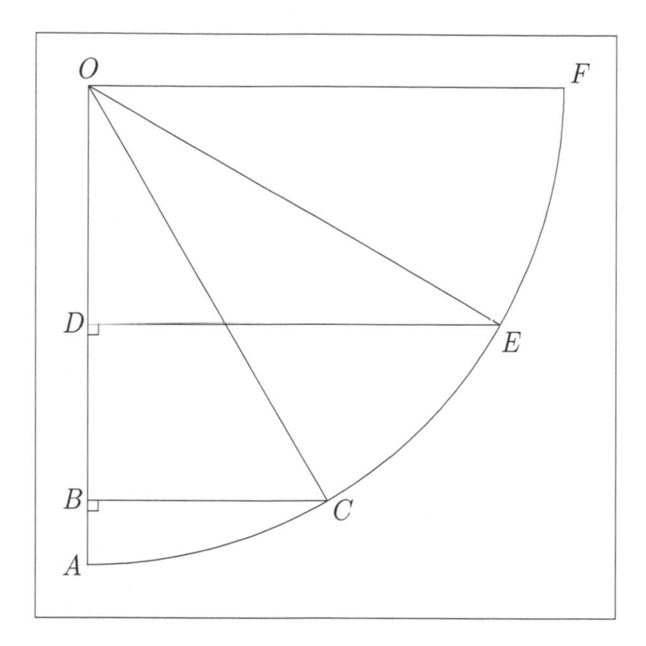

Figure 14: The rising arcs and the ecliptic

in fact, better than we found, which could imply that he used a better method for computing it or took the number from another source.

The numbers corresponding to the remaining six signs can easily be inferred.

($42^{a–b}$) **[If, in three separate right triangles,] the Sines of the endpoints of the [first] three signs are the hypotenuses, which are on the ecliptic, and the Sines of the [corresponding] declinations [of the sun] are the legs, which are on the respective diurnal circles, [then] the rising arcs [on the celestial equator] are [given] by the corresponding uprights, which are on the [corresponding] diurnal circle.**

The rising arcs determined here are the segments of the celestial equator corresponding to the first three signs on the ecliptic (measured from the vernal equinox). Since segments of the equator correspond to time, these arcs determine the rising times of the first three signs.

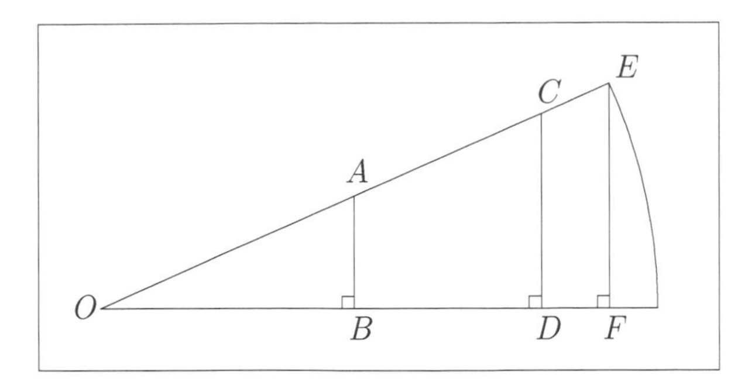

Figure 15: The rising arcs, the ecliptic, and the equator

In Figure 14, the arc $ACEF$ is part of the ecliptic, the point A marking the vernal equinox. The angles AOC, COE, and EOF are each 30°.

In Figure 15, the line $OACE$ is the ecliptic and the line $OBDF$ is the celestial equator. The point O marks the vernal equinox. The points O, A, C, and E are separated by 30°. Therefore, in Figure 15 the lengths of OA, OC, and OE correspond, respectively, to the lengths of BC, DE, and OF in Figure 14. It follows that in Figure 15, the line OA is Sin(30°), the line OC is Sin(60°), and the line OE is Sin(90°). Furthermore, as is easily seen, the line AB is Sin(δ_1), the line CD is Sin(δ_2), and the line EF is Sin(δ_3), where δ_1, δ_2, and δ_3 are as defined in the preceding.

From all this we see that the line OB in Figure 15 is the Sine of the arc on the celestial equator corresponding to the arc OA on the ecliptic, and similarly for OD and OC, and OF and OE. By subtracting appropriately, i.e., subtracting $|OB|$ from $|OD|$ and $|OD|$ from $|OF|$, we can find the Sines of the equatorial rising arcs of the first three signs.

(42^{c-d}–43) **The ascendant [can be found as follows]. The *pala*s of the given *ghaṭikā*s [i.e., the given time in *pala*s rather than in *ghaṭikā*s] are diminished by the remaining part of the rising time of the sign that the sun is in [i.e., the time it will take until the sun enters the next sign] and the *pala*s [of the rising times] of the signs that have risen completely. The result of the division of the product of the result and 30 by the rising time of the**

sign that is currently rising is in degrees and so on. [This quantity] increased by the signs preceding the currently rising sign beginning from Aries and diminished by the degrees of precession is the [longitude of the] ascendant.

The time [elapsed since sunrise] in *pala*s and so on [can be found] from the difference between [the longitude of] the precessional sun and [the longitude of] the precessional ascendant.

When the ascendant is to be computed at night, the time is [found] from [the longitude of] the sun increased by six signs.

[When] both [the sun and the ascendant] are in the same sign, the rising time [of that sign] is multiplied by the degrees [of the angular distance] between them and divided by 30.

I have taken *pūrva* in 42$^{\text{c–d}}$–43 in the sense of "signs that have risen completely." Literally, however, *pūrva* means "preceding," the sense in which the word is used in 44. But taking the meaning to be the "preceding signs" does not make sense here, whereas "signs that have risen completely" fits the context.

The ascendant is the point of the ecliptic which is rising on the local horizon at a given time. The procedure for computing the ascendant is illustrated in Figure 16. In the figure, the horizontal line is the local horizon and the oblique line is the ecliptic. The ecliptic is divided into signs by small perpendicular line segments. Suppose that the sun is at the position P in the sign S_1, and let A denote the ascendant. Of the signs following S_1, S_2 and S_3 have risen completely above the horizon, while S_4 is rising.

Let the rising times of S_1, S_2, S_3, and S_4 be t_1, t_2, t_3, and t_4, respectively, and let us assume that t *pala*s have elapsed since sunrise. At sunrise, the sun and the ascendant coincided, so between sunrise and the given time, a portion of S_1, all of S_2 and S_3, and a portion of S_4 have risen.

If the angular distance between the sun and the beginning of S_2 is a, it took this segment $\frac{a \cdot t_1}{30}$ *pala*s to rise. It further took $t_2 + t_3$ *pala*s for S_2 and S_3 to rise. Subtracting these times from

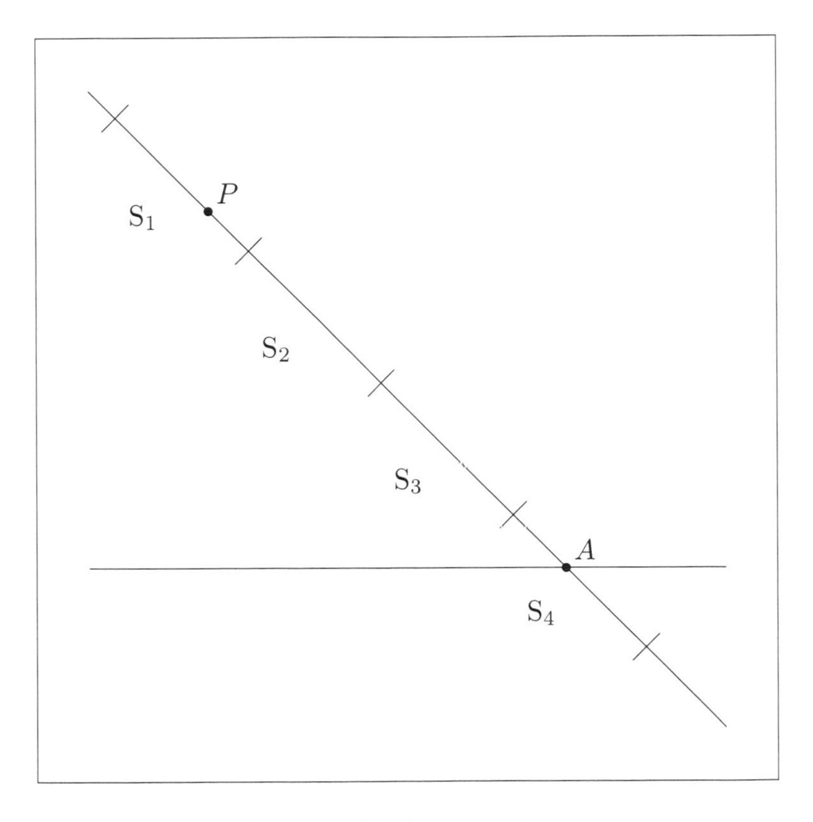

Figure 16: The ascendant

t, we get

$$s = t - \frac{a \cdot t_1}{30} - t_2 - t_3 \tag{123}$$

as the *pala*s that it has taken the segment of S_4 from its beginning to the point on the horizon to rise. If this segment spans the angular distance b, then

$$b = \frac{30 \cdot s}{t_4}. \tag{124}$$

We now know the point of S_4, i.e., A, that is on the horizon. This point is the ascendant. To find its proper longitude, we first have to add all of the preceding signs beginning with Aries, and then convert this tropical longitude to a sidereal one by subtracting the degrees of precession.

(44) **Aries and so on move eastward away from the celestial equator when the *pravaha* wind is diminished, and westward when [the Pravaha wind] is increasing. Therefore, the degrees of precession are not the same each year.**

In the following we will continue to use the word "precession" for the motion of the equinoxes against the backdrop of the fixed stars.

In the preceding, we have translated the Sanskrit word *ayana* as precession and the word *sāyana* as "precessional," which is useful for immediate understanding. However, Jñānarāja's model does not actually operate with precession of the equinoxes, but rather with trepidation of the equinoxes. Trepidation of the equinoxes is the theory that the vernal equinox moves a certain angular distance to the east, then stops and moves the same distance to the west, then stops and moves the same distance to the east, and so on *ad infinitum*.[208]

No details regarding the rate of the motion are given by Jñānarāja. More details are given in the *Sūryasiddhānta*, where it is stated that the trepidation of the equinox extends over an arc of 27° at a rate of 54″ per year.[209]

[208] For the history of precession and trepidation in early India, see Pingree 1972.

[209] *Sūryasiddhānta* 3.9–11. See Pingree 1978a, 610.

(45) **The difference between the degrees of [the longitude of] the sun [determined] from the noon shadow and [the longitude of] the sun [determined] from a *karaṇa* [work] is the number of degrees of precession.**

In this regard, after a year, a difference in minutes of arc [accumulated between the tropical and sidereal longitudes of the sun] is seen directly.

For the sake of computation of the ascendant, the ascensional difference, and the declination, [this difference must be taken into account].

From the shadow cast by a gnomon at noon, we can find the declination of the sun, and from the declination its tropical longitude (see 2.1.27–29 and commentary for how to find the declination and longitude). This will differ from the longitude found by carrying out the computations prescribed in a *karaṇa* treatise (a type of astronomical treatise that provides formulae, but not theory; see the Introduction, page 19), which will give the sidereal longitude. The difference between the two longitudes is the degrees of precession.

As the precessional rate is given as 54″ in the *Sūryasiddhānta*, it is roughly correct that one should see a difference of about a minute of arc after a year.

(46) **A day [the daylight period of which is] measured by 30 *ghaṭikās* is to be observed. The difference between the degrees of [the longitude of] the sun on that day and [the longitude of the sun] on the equinoctial day is the number of degrees of precession. The computation of the declination, the ascendant, and the ascensional difference is to be done from [the solar longitude] corrected by these [degrees of precession].**

A day whose daylight period is 30 *ghaṭikās* is an equinoctial day. The idea seems to be that you compute the equinoctial day by an algorithm relying on an outdated value for current precession, and then correct the amount by observing the difference between that and the true equinoctial day with daylight equal to 30 *ghaṭikās*.

(47) **The *ghaṭikās* corresponding to the difference between the correct computation of a shadow [as explained] by the primeval sages and the visible shadow is due to the [precessional] difference.**

The method recorded here, which was enunciated by the ancients, is easy, as it is based on proportions and so on, and it does not give errors.

A shadow computation refers to using the shadow cast by a gnomon for determining various astronomical values. This will be taken up in detail in the next section. Now, when one comes across a shadow computation in a work that differs from what we see in practice, but is otherwise correct, it is to be understood that the difference is due to precession.

(48) **Thus is the planetary rectification produced from the velocity in the beautiful and abundant *tantra* composed by Jñānarāja, the son of Nāganātha, which is the foundation of [any] library.**

9. Chapter on Mathematical Astronomy

Section 3: *Three Questions (on Diurnal Motion)*

This section deals with questions pertaining to diurnal motion. The "three" in the title refers to direction, place, and time.

(1) **For the sake of computing the results caused by direction, place, and time, I will now present the section entitled "Three Questions" in the *Siddhāntasundara*, which has a collection of flowers [in the form of] good *vāsanā*s [that is, methods], which is a moving and very wide spherical tree, and which is the greatest.**

The word *golataroḥ* (literally, sphere-tree) could also refer to the earth, but it seems most likely that it is this treatise that has flowers in the form of *vāsanā*s.

(2) **The shadow of a gnomon that is straight and positioned on ground that has been made even by means of water [falls along the] south-north [line] at [the time of] the *ghaṭikā*s of midday. The east and west directions are produced from the tail and head of [a figure in the shape of] a fish produced from it [the north-south line].**

This verse is identical to verse 2.1.26, at which place comments are given.

(3) **After placing a flexible and straight rod on a peg at the center of a circle [drawn] on even ground on a raised platform, so that it is pointing toward [the rising point of] the sun, the eastern direction is [found] by means of the degrees in the arc produced by the Sine of the**

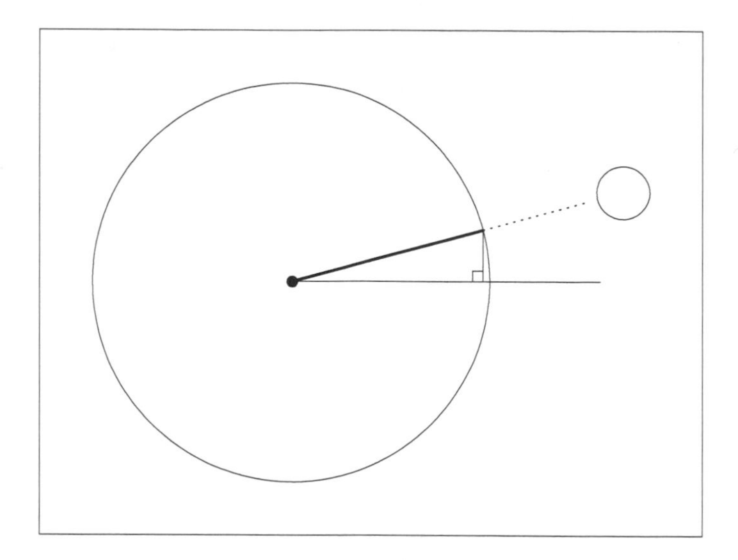

Figure 17: The east-west line

amplitude placed in the opposite direction from its [the rod's] tip on the circumference [of the circle].

Consider Figure 17. The larger circle is the circle described in the verse, and the smaller circle is the sun. The rod, which is the bold line, is oriented toward the position of the rising sun.

Now, unless it is the equinoctial day, the sun will not rise exactly due east. The angle between due east and the rising point of the sun is called the rising amplitude, or simply the amplitude. If the amplitude, which can be found through computation, is known, we can find due east on the circle; on the figure it is the second, non-bold line emanating from the center of the larger circle.

The reason that the amplitude is placed in its opposite direction is that this arc is normally measured from the eastern point, but we are extending it from the rising point of the sun.

(4) **The western direction extending from the said center is the shadow from a peg at the center. On account of the sun ... [?]. The western and eastern directions are [determined] in that manner in order. The northern**

[direction] is given by the polestar. Such is the determination of the cardinal directions.

This verse deals with the establishment of the remaining directions. The procedure is clear from the translation. Note, however, that the meaning of the passage (*ravivaśena viśaty apaiti*) is unclear and this passage might be corrupt.

(5) **When the precessional sun is situated at the end of Virgo or at the end of Pisces, the shadow [cast by a gnomon] at noon is the equinoctial shadow.**

[In a right-angled triangle,] this [equinoctial shadow] is the leg, the gnomon is the upright, and the hypotenuse is the equinoctial shadow.

This is the first figure among the figures arising from the latitude.

On the equinoctial day, that is, when the sun is at one of the equinoctial points (the two intersections between the ecliptic and the celestial equator) and thus on the celestial equator, the shadow cast by the gnomon at noon is called the equinoctial shadow. The corresponding hypotenuse (that is, the distance between the top of the gnomon and the end of the shadow) is called the equinoctial hypotenuse. It should be noted that the gnomon is always considered to have the height 12.

This is illustrated in Figure 18, where g is the gnomon (of height 12), s_0 the equinoctial shadow, and h_0 the corresponding hypotenuse, which is called the equinoctial hypotenuse. Note that the lengths are chosen so as to conform with the situation at the latitude of Pārthapura.

Now, since the sun is on the celestial equator on the equinoctial day, the angle between the gnomon and the equinoctial hypotenuse is equal to to the local terrestrial latitude ϕ. Similarly, the angle between the equinoctial shadow and the equinoctial hypotenuse is the local co-latitude $\bar{\phi}$ (the co-latitude is defined as 90° diminished by the latitude, that is, $\bar{\phi} = 90° - \phi$).

The triangle given here is the first in a series of similar triangles that will be presented in the following.

(6–11) **The vertical line from the equinoctial point [at**

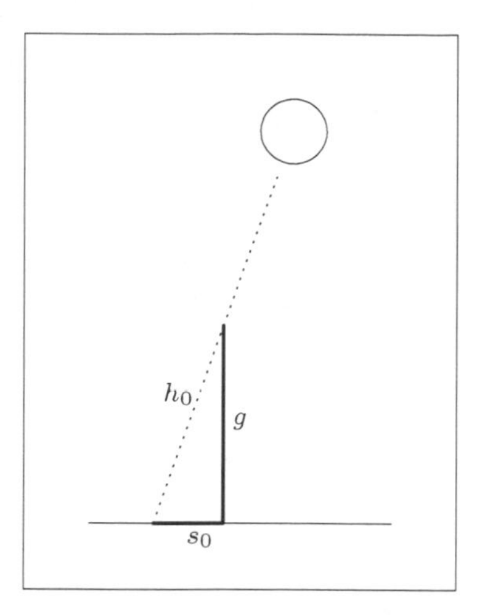

Figure 18: Shadow cast by a gnomon on an equinoctial day at noon

noon] to the earth is the Sine of the local latitude, which is the leg [of a right-angled triangle]; the radius, which is between the center of the earth and the equinoctial point, is the hypotenuse; and the Sine of the co-latitude is said to be the upright. This [upright] is situated at the tip of the Sine of the local latitude from the center of the earth.

[In a right-angled triangle,] the earth Sine, which is known from the previous, is the leg. The [corresponding] upright is the Sine of the declination [of the sun], and the hypotenuse is the Sine of the amplitude.

[This right-angled triangle can be divided into two right-angled triangles as follows.] One leg is the earth Sine, another the Sine of the declination [of the sun]. The base is the Sine of the amplitude, and the upright is the Sine of the six-o'clock altitude.

[In the first of these triangles,] the first segment of the base is the part beginning at the tip of the Sine of the amplitude; it is the upright. The [corresponding] hypotenuse is the Sine of the declination, and the leg is the Sine of the six-o'clock altitude.

[In the second triangle,] the other segment of the base is the leg. The [corresponding] upright is the Sine

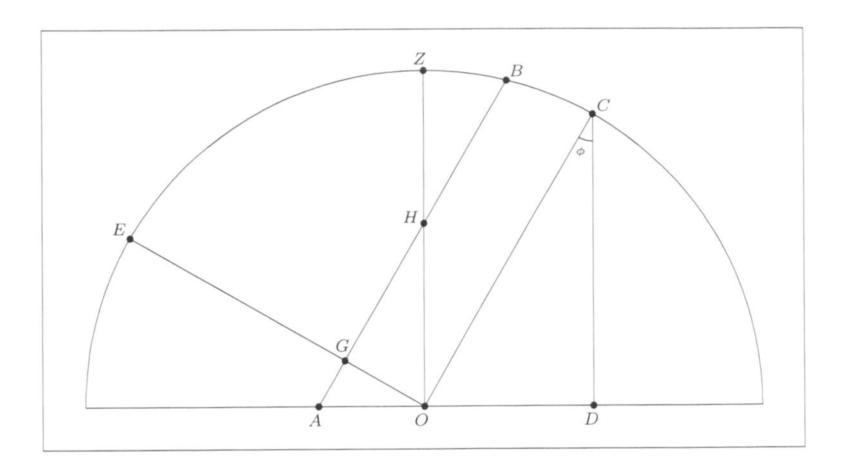

Figure 19: Similar right-angled triangles via analemma

of the six-o'clock altitude, and the hypotenuse in this right-angled triangle is the earth Sine.

The perpendicular from the [position of] the sun when it is on the prime vertical [that is, the Sine of the altitude of the sun when it is on the prime vertical] is called the Sine of the prime-vertical altitude. This as the upright, the Sine of the amplitude as the leg, and the prime-vertical hypotenuse as the hypotenuse [form] another right-angled triangle.

Another right-angled triangle is [formed when] the upright is the Sine of the prime-vertical altitude diminished by the Sine of the six-o'clock altitude, the hypotenuse is the prime-vertical hypotenuse diminished by the earth Sine, and the leg is the first segment of the Sine of the amplitude.

Both the Sine of the prime-vertical altitude and the prime-vertical hypotenuse have two segments.

[In another right-angled triangle,] the leg is the Sine of the declination, and the upright is the prime-vertical hypotenuse diminished by the earth Sine.

When the sun is situated north or south of the prime vertical, the hypotenuse is in this case the Sine of the prime-vertical altitude [?].

Figure 19 shows the analemma from which the similar right-angled triangles given by Jñānarāja are derived. The horizontal line is the local horizon, and the half circle is the part of the local meridian that is above the horizon. The line OC is the equator.

The six-o'clock circle is the great circle that is perpendicular to the celestial equator and passes through the east and west points (it is called the six-o'clock circle because the sun crosses it at 6 in the morning and again at 6 in the evening); it is the line OE in Figure 19. The diurnal circle is the circle describing the diurnal motion of the sun (at the given time), that is, the path of the sun's apparent motion around the earth (following the apparent motion of the fixed stars); it is the line AB in Figure 19. The prime vertical is the great circle that is perpendicular to the horizon and passes through the east and west points; it is the line OZ in Figure 19.

The six-o'clock altitude is the altitude of the point of intersection between the six-o'clock circle and the diurnal circle.

The earth Sine was introduced in the previous section (see *Siddhāntasundara* 2.2.34). It is the arc of the diurnal circle between the horizon and the six-o'clock circle, that is, the angular distance traveled by the sun between its rising and its reaching the six-o'clock circle. Of course, this is the same angular distance as between the sun reaching the six-o'clock circle in the evening and its setting. Taken either way, it is the line AG in Figure 19.

The amplitude is the arc on the horizon between the east point (O in Figure 19) and the intersection between the horizon and the six-o'clock circle (the point A in Figure 19). Note that we could similarly have defined the amplitude with respect to the west point. With our definition, it is the angular distance between the east point (O) and the point where the sun rises or sets. It is the line OA in Figure 19.

Note that the celestial radius used in the above is taken to be equal to the trigonometric radius for simplicity of computation. As such, the length of OC in Figure 19 is $R = 3{,}438$. We can now use Sines to express other lengths. For example, the length of the line OG is thus the Sine of the declination of the sun, that is, $\mathrm{Sin}(\delta)$.

Figure 20 shows triangle OAH in Figure 19 with further divisions according to the verses. The angle ϕ is the local latitude, and the angle $\bar{\phi}$ is the local co-latitude.

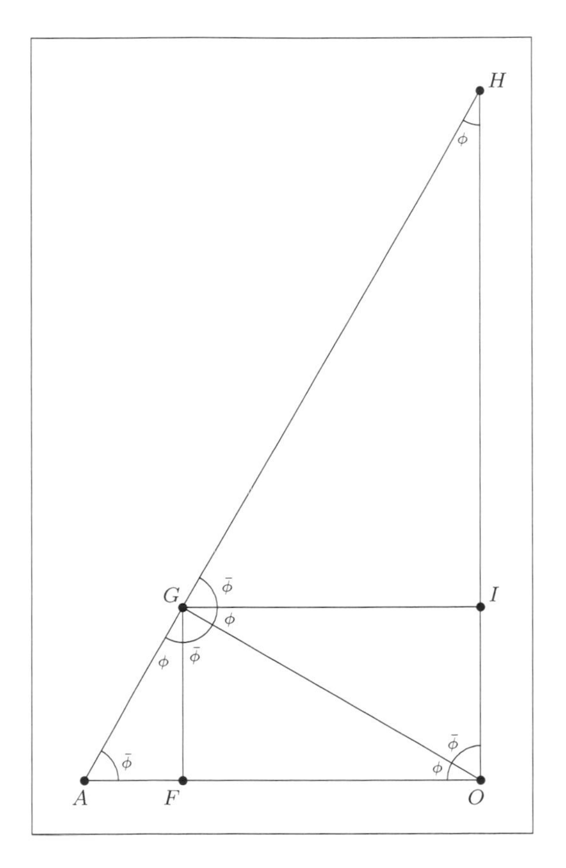

Figure 20: Triangle OAH from Figure 19

The eight triangles (given in the order: leg, upright, and hypotenuse) listed by Jñānarāja are as follows:

1. the equinoctial shadow, the gnomon, and the equinoctial hypotenuse, which is shown in Figure 18;

2. triangle OCD in Figure 19, containing the Sine of the latitude OD, the Sine of the co-latitude CD, and the radius OC;

3. triangle OAG in Figure 20, containing the earth Sine AG, the Sine of the sun's declination OG, and the Sine of the amplitude OA;

4. triangle OFG in Figure 20, containing the Sine of the six-o'clock altitude GF, the first segment of the Sine of the amplitude OF, and the Sine of the declination OG;

5. triangle AFG in Figure 20, containing the second segment of the Sine of the amplitude AF, the Sine of the six-o'clock altitude FG, and the earth Sine AG;

6. triangle OAH in Figure 20, containing the Sine of the amplitude OA, the Sine of the prime-vertical altitude OH, and the prime-vertical hypotenuse AH;

7. triangle GHI in Figure 20, containing the first segment of the Sine of the amplitude GI, the Sine of the prime-vertical altitude diminished by the Sine of the six-o'clock altitude HI, and the prime-vertical hypotenuse diminished by the earth Sine GH; and

8. triangle OGH in Figure 20, containing the Sine of the declination OG, the prime-vertical hypotenuse diminished by the earth Sine GH, and the Sine of the prime-vertical altitude OH.

That all of these triangles are similar is easy to see.

(12^{a-b}) **By means of the measures of the leg, hypotenuse, and upright in any one among these [right-angled triangles], a computation of the leg and so on in another**

right-angled triangle [among the ones given] can be carried out by means of a proportion.

Since all the right-angled triangles given are similar, we can use a proportion to determine unknown sides. Jñānarāja will give some examples of this below in verses 13–16.

(12^{c-d}) **The hypotenuse [in one of the right-angled triangles] is [equal to] the square root of the sum of the square of the leg and the square of the upright. The upright is [equal to] the square root of the difference of the square of that [hypotenuse] and the square of the leg.**

If l, u, and h are the leg, the upright, and the hypotenuse, respectively, in a right-angled triangle,

$$h = \sqrt{l^2 + u^2}, \tag{125}$$

and

$$u = \sqrt{h^2 - l^2}. \tag{126}$$

Both results follow directly from the Pythagorean theorem.

(13) **If the upright and the leg are, respectively, the gnomon and the equinoctial shadow when the hypotenuse is the equinoctial shadow, then what are they when it [the hypotenuse] is measured by the radius? [In this case,] the upright and the leg are the Sine of the colatitude and the Sine of the latitude, respectively.**

This verse provides an example of how to use a proportion based that is on two of the given triangles, more specifically the first and second triangles found in the list of similar triangles on page 223.

If l and u denote, respectively, the leg and the upright of the second triangle, then it follows from two proportions that

$$\frac{l}{R} = \frac{s_0}{h_0} \tag{127}$$

and

$$\frac{u}{R} = \frac{g}{h_0}, \tag{128}$$

from which it is seen that $l = \mathrm{Sin}(\phi)$ and $u = \mathrm{Sin}(\bar{\phi})$, which is what the verse states.

(14) **[If] the leg [of a right-angled triangle] is the shadow when the shadow hypotenuse is the one at noon, then what is it [the leg] when the hypotenuse is the radius? The Sine of the degrees of the zenith distance [of the sun] is attained as the result. From this [Sine] the degrees of the latitude are to be computed as was explained.**

Consider Figure 6 and the commentary on verses 2.1.27–29[a]. On an arbitrary day, the shadow triangle (gnomon, shadow, and shadow hypotenuse) is similar to the right-angled triangle with hypotenuse OS and one leg along the line OZ in Figure 6.

(15–16) **The equinoctial shadow is [separately] multiplied by the Sine of the six-o'clock altitude, the Sine of the amplitude, the first segment of the Sine of the amplitude, and the second segment of the Sine of the amplitude, and [each result is] divided [separately] by the equinoctial shadow. [The results] are the Sine of the declination of the sun, the prime-vertical hypotenuse, the upper segment of that [prime-vertical hypotenuse], and the earth Sine.**

The Sine of the six-o'clock altitude, the Sine of the amplitude, the first segment of the Sine of the amplitude, and the second segment of the Sine of the amplitude are [separately] multiplied by 12 [that is, the height of the gnomon] and divided by the equinoctial shadow. [The results] are, respectively, the first segment of the Sine of the amplitude, the Sine of the prime-vertical altitude, the upper segment of that [Sine of the prime-vertical altitude], and the Sine of the six-o'clock altitude.

The first four proportions given are derived from the similarity of the first triangle with the fourth, the sixth, the seventh, and the fifth triangles, respectively, from the list of similar triangles on page 223. The next four proportions are derived from the similarity of the first triangle with the same four triangles in the same order.

Now [some] questions concerning location.

Jñānarāja now turns his attention to direction. His discussion begins by posing a problem.

(17) **[Narrative translation:] Tell [me] the length of the journey of the swift man, who, upon learning that his friend had gained kingship and was sitting on the lion's seat [that is, the throne], deprived of luster went north [to a place where] he had the luster of a former king.**

[Technical translation:] Tell [me] the length of the journey of the swift man, who, upon learning that the sun had attained lordship sitting in the lion's seat [that is, was in Leo], cast no shadow and who went north [until] he had a shadow of [length] 16 [falling] toward the east.

This is the first of the problems posed in this section where the poetic technique of double entendre (*śleṣa*) is employed. As was stated in the introduction (page 36), it is necessary to give two separate translations for each of these verses, one narrative and one technical.

As can be seen, the narrative translation sets the scene. A king has been ousted by his friend, who now occupies the throne, and flees to another region, where he is given recognition as a former king. A question is posed, but the narrative does not give the information needed to answer it. For this we need to reread the verse, leading to the technical translation.

The problem can be solved as follows. We have two places: P_1, the point of departure, and P_2, the point of arrival. As the journey is due north and the circumference of the earth is 5,059 *yojanas* according to the *Siddhāntasundara* (see 1.1.74), we only need to know the latitudes of P_1 and P_2, ϕ_{P_1} and ϕ_{P_2}, to determine the distance between them.

At P_1, there is no shadow, so the sun is in the zenith, which means that $\phi_{P_1} = \delta$, where δ is the declination of the sun. Now, the statement that the sun is in the lion's seat means that the sun is in Leo. In fact, Jñānarāja intends for it to be at the midpoint of Leo (see solution to this verse below), in other words, we know that λ^*, the sun's tropical longitude, is $\lambda^* = 4^s 15° = 135°$ (the notation 1^s means 1 sign, that is, 30°). By using Jñānarāja's small Sine table and the formula $\text{Sin}^{\ell}(\delta) = \text{Sin}^{\ell}(\varepsilon) \times \text{Sin}^{\ell}(\lambda)/R$, where $\varepsilon = 24°$ is the obliquity of the ecliptic, we get $\text{Sin}^{\ell}(\delta) =$ 45;2, and then $\delta = 16°31'$ (see (107)). Hence, $\phi_{\text{P}_1} = 16°31'$.

At P_2, the shadow has length 16 and falls due east, which means that the sun is on the prime vertical. We can find α, the altitude of the sun, by the proportion $\frac{\alpha}{R} = \frac{g}{\sqrt{g^2+s^2}}$, where g is the length of the gnomon, that is, 12, and s is the length of the shadow, that is, 16, which gives $a = 96$. Since the sun is on the prime vertical, $\frac{\text{Sin}(\delta)}{\alpha} = \frac{\text{Sin}(\phi_{\text{P}_2})}{R}$, which gives us that $\text{Sin}(\phi_{\text{P}_2}) = 75\frac{3}{60}$ and $\phi_{\text{P}_2} = 28°19'$.

Now, the degrees that the man traveled north are $\phi_{\text{P}_2} - \phi_{\text{P}_1} = 28°19' - 16°31' = 11°48'$, which by the next verse corresponds to $14 \cdot 11;48$ *yojanas*, or 165;12 *yojanas*. This is approximately 1,300 kilometers, roughly the distance between Hyderabad and Delhi.

The term *siṃhāsana* (lion's seat) used in the verse is an astrological term found in the *Bṛhatpārāśarahorāśāstra*.[210] However, it is not used in this sense here.

(18) [Here is] the rule: The difference of the latitudes at two locations is multiplied by 14. Thus, the *yojanas* of the journey [are found].

As the circumference of the earth is 5,059 *yojanas* (see verse 1.1.74), there are

$$\frac{5{,}059}{360} = 14\frac{19}{360} \approx 14 \tag{129}$$

yojanas per degree on any great circle through the poles. Therefore, if a person is traveling straight north or straight south, the length of his journey in *yojanas* can be found by multiplying the difference of the latitude of the place from which he set out and

[210] *Bṛhatpārāśarahorāśāstra* 6.45–47 (see Santhanam 1984, 1.87). I am thankful to Martin Gansten for calling my attention to this astrological usage of the term *siṃhāsana*.

the latitude of his destination by 14. For this formula, see also *Siddhāntasundara* 1.1.25.

In this case, the sun is at the midpoint of Leo [that is, its tropical longitude is] $4^s15°$. **The first shadow [has length]** 0; **the second [has length]** 16 **[falling toward] east. In this case, by the method given in verse 22, having computed the degrees of the latitude from the equinoctial shadow, the *yojanas* found are** 167.

Most manuscripts give 157 *yojanas* rather than 167. However, we got 165;12 *yojanas* above. Note that Jñānarāja's solution is really only a sketch.

(19) **[Narrative translation:] When a good goose perished in the mouth of a crocodile, the best among sages left [lake] Mānasa. [While on his way,] he saw a light in the northeastern direction resembling [the deity] Śiva. Tell me the equinoctial shadow at that place!**

[Technical translation:] When the sun, which was at the beginning of Capricorn, was setting, the best among sages left [lake] Mānasa. [While on his way,] he saw that the shadow [of a gnomon] was [of length] 11 **in the northeastern direction. Tell me the equinoctial shadow at that place!**

Both the narrative and technical meanings of the verse are clear, but the solution to the problem, as per the following verses, is not clear.

(20–21) **The difference between the square of the Sine of the declination [of the sun] and the square of the Sine of the altitude [of the sun] is multiplied by** 2. **This is the divisor [*hāra*]. It is multiplied by the sixth part of the Sine of the altitude [of the sun]. This is the mean leg [*bāhu*]. Whatever is the difference between the square of the Sine of the declination [of the sun] and [this] leg multiplied by** 2 **times the divisor, the square root of the difference or the sum of the square of that and the square of the mean [leg] is decreased or increased by the**

mean [leg], depending on whether [the Sine of] the altitude [of the sun] is greater or smaller than [the Sine of] the declination. In this case, whatever is the result [from the division of this quantity] by the divisor, that is the equinoctial shadow. [This] multiplied by the direction Sine is the [given?] shadow. The [mean?] leg divided by the [corresponding shadow-]hypotenuse [?] is the given direction.

When the direction Sine [is found] by means of the radius R, what [is found] from the [given?] shadow? When it is understood that the leg is not subtracted from that which is to be subtracted [?], then, by means of an inverse [procedure] that which is to be subtracted is entire by means of one's own intelligence.

The direction Sine is the Sine of the degrees separating the current altitude circle of the sun from the prime vertical.

The procedure in these two verses is not clear.

The demonstration verse for verses 20–21 [is given now].

Jñānarāja now gives the demonstration (*vāsanā*) for verses 20–21 above.

(22) **One should compute the base of [the Sine of] the altitude by means of the unknown equinoctial shadow. [The sine of] the amplitude [is found] from the correction of the leg by that [base of the Sine of the altitude]. The square of the *palabhā* [equinoctial shadow?] is increased by the square of 12 [that is, 144]. [This is] the square of the equinoctial hypotenuse. The square of [the Sine of] the amplitude is divided by the square of 12 multiplied by the square of the Sine of the declination [of the sun]. The square of the previous [eastern?] [Sine of] the amplitude is equal to that. In this case, the equinoctial shadow [is found] from an application of *bījas* from the same *karaṇa* [treatise].**

The procedure outlined here is unclear.

[The tropical longitude of] the sun is $9^s0°0'0''$. By means of the method [for computing a sine using the table] of small Sines, the Sine of the declination [of the sun] corresponding to that [tropical longitude] is 64;20, [the Sine of] the altitude is 117;48, and the direction Sine is 112;0. The equinoctial shadow found [from this] is 1;7, and the [corresponding] degrees of terrestrial latitude are 5;7.

This verse provides the solution to the problem posed in verse 19.

Since the tropical longitude of the sun is $\lambda^* = 9^s$, $\text{Sin}^\ell(\lambda^*) = R$ (note that since we are operating with small Sines, $R = 160$). By the formula given in *Siddhāntasundara* 2.2.32$^{\text{c–d}}$,

$$\text{Sin}^\ell(\delta) = \frac{\text{Sin}^\ell(\lambda^*) \cdot \text{Sin}^\ell(\varepsilon)}{R} = \text{Sin}^\ell(\varepsilon) = \text{Sin}^\ell(24°). \qquad (130)$$

Using the formula given in *Siddhāntasundara* 2.2.13–14, we get

$$\text{Sin}^\ell(24°) = \frac{6}{9} \cdot 23 + 24 + 25 = 64;20. \qquad (131)$$

The rest of the solution is not clear.

(23) **[Narrative translation:] The friend, having the splendor of the previous [deity] Indra, causes pain to the hands of a virgin, his hands being harsh hands; [he is] clad in good cloth and jewels, located amongst the star-maidens [?], accompanied by his teachers and knowledgeable poets, and being the lord of chariots, horses, and men [probably referring to an army regiment]. O friend, tell me by how many *yojana*s the city where [this man is] is from [the city of] Dhārā in the direction of southeast.**

[Technical translation:] The sun, the sharp-rayed one, which is causing pain to Virgo with its rays, throwing a shadow equal [in legth] to 14 [falling] toward the east. [Rest of the verse unclear.]

Neither the narrative nor the technical translation is clear, especially the technical translation. The approach to a solution, outlined in verses 24–30, is also not clear.

(24) **The direction Sine is multiplied by 12 and divided by the radius and by the local equinoctial shadow. [That is] the unknown. The product of [the Sine of] the amplitude and 12 is divided by the equinoctial shadow. That is the known. Then [that] is mean [?]. There is a pair of the unknown and the known which is a product [?]. Whatever is the square root of the square of the unknown increased by 1, [that] is the divisor, it is said. The mean divided by the divisor and [then] halved is the [desired] result.**

The procedure in the verse is not clear.

(25) **The difference between the square produced from the known and the square produced from the half-diameter [?] is to be added to the result [?] that has been squared. When the known is less than the half diameter or greater than it, the difference of the two [quantities] is to be computed. The square root of that divided by the divisor increased or diminished by the result is the before-mentioned Sine of the zenith distance [of the sun]. When multiplied by the degrees in its arc, its measure is 14. These are the *yojana*s in the distance [between two locations on the same meridian].**

The procedure is not clear.

(26) **The equinoctial hypotenuse at a city [whose latitude is] known is multiplied by the Sine of the latitude of a city [whose latitude] is not known and divided by 12. This is [the Sine of] the amplitude. It is said that by means of [the Sine of] the amplitude of the sun the Sine of the *yojana*s in the distance [between the two locations can be found]. If it is the given direction, it is [the Sine of] the zenith distance [?].**

The procedure is not clear.

In this case, [the tropical longitude of] the sun is $5^s10°0'0''$**.**

The shadow is 14 and the [shadow] hypotenuse is 18;26. By means of [the method of] small Sines, [the Sine of] the altitude is 104 and the Sine of the declination is 22;0. The equinoctial shadow is 2;37 and the equinoctial hypotenuse is 12;17. [The Sine of] the zenith distance, attained as was explained, is 118. The *yojanas* are 667.

Jñānarāja here gives the solution to the problem posed in verse 23.

It appears that the *yojanas* between the two cities in verse 23 are the number of degrees in the zenith distance multiplied by 14. Let this zenith distance be z. If $\mathrm{Sin}^{\ell}(z) = 118$, then $14 \cdot z \approx 665$. Verse 23 is still not entirely clear.

(27) **Supposing that the sphere of the earth is a perfect sphere, one's own city is at an intersection with the sky [?]. Therefore, it is to be imagined that the given city is on a circle through the zenith and the sun [?] located at the degrees of the sky [?]. In this case, [the Sine of] the zenith distance is the Cosine of that by means of the degrees between the two cities. [The Sine of] the altitude is the leg [of a right-angled triangle]. The degrees between that and the prime vertical is the base of [the Sine of] the altitude; it is [the Sine of] the amplitude.**

The procedure is not clear.

(28) **[If] the direction Sine [is found] by means of the Sine of the degrees of the distance [between two locations?] being [equal to] the radius *R*, what leg [of a triangle is found] by means of the Sine of the degrees of the distance being [equal to] the unknown? In this case is [the Sine of] the amplitude diminished or increased by the leg measured by the result. This is the base of [the Sine of] the altitude in the Northern and Southern Hemispheres, respectively.**

The procedure is not clear.

(29) **If the upright measured by [the longitude of] the sun [is found] from one's own equinoctial shadow, then what Sine of the upright is the result in case of the leg measured by the base of [the Sine of] the altitude?**

The square of 3 **[or of the radius** R?**] is diminished by the square of the Sine of the leg which is not known. [The result] is the square of the Cosine of the degrees of the distance. Eastern [part] of the sky [?].**

The procedure described is not clear.

(30) **It is said that it is the same by means of a square [?]. For the sake of the Sine of the arc of the degrees of the distance, a *bīja* correction is to be applied for the sake of making them equal.**

The procedure is not clear.

For *bīja* corrections, see the note in the commentary on *Siddhāntasundara* 2.1.83–84.

(31) **[Narrative translation:] When the good friend [that is, Lakṣmaṇa], resting on the lap-part of Rāma, perished, Hanumat went from Laṅkā toward the northeastern direction with a desire for the cure for the arrow-wound. He stood gazing at the brightness of the straw similar to rocks that reached up to the peak of Mt. Śrīkaṇṭha. O learned one, tell the journey if you are the sun illuminating the digits of the moon.**

[Technical translation:] When the sun, situated in the 9th degree of Virgo, set, Hanumat went from Laṅkā toward the northeastern direction with a desire for the cure for the arrow-wound. He stood gazing at the noon shadow equal [in length] to 7 and facing the cardinal point of Siva's direction [northeast]. O learned one, tell the journey if you are the sun illuminating the digits of the moon.

The verse is not entirely clear.

(32) **[Narrative translation:] By means of which time**

do I see a friend being seated on the lion's seat [that is, the throne], obeying the command of [the god] Indra by means of enjoyment on the same circle [?], and bowing down to the foot of [the deity] Viṣṇu in a known city?

[Technical translation:] By what time do I see the sun being in Leo, filling [?] the eastern direction by means of motion on the prime vertical, and bowing down to the foot of Viṣṇu [a star?] in a known city.

Neither the narrative nor the technical parts of the verse are clear.

(33) **[The Sine of] the zenith distance [of the sun] is to be computed, as explained, by means of [the Sine of] the amplitude of the sun. By means of that [are found] the *ghaṭikās* in the given direction-shadow [?]. Entering the given direction in these *ghaṭikās*, the shadow of the sun [can be found], and from that the given direction is to be computed.**

(34) **Tell me the shadow of the sun extending its part in the direction of northeast for one gone to Viṣṇu [?] in the vicinity of Ātmatīrtha on the bank of the [river] Godāvarī whose equinoctial shadow is** 4;20. **Tell [me] also, O friend, the equinoctial shadow measured by** 5;40 **in the direction of northeast, by how many *yojana*s by one bringing [?]? I desire Vārāṇasī!**

Neither the narrative part nor the technical part of this verse is fully clear.

Ātmatīrtha is a sacred place near Jñānarāja's hometown of Pārthapura. Vārāṇasī, also known and Kāśī, and earlier as Benares or Banaras, is a city in the Indian state of Uttar Pradesh. It is considered holy city in Hinduism.

Here [the longitude of] the sun is 4;15. **The Sine of the declination [of the sun] is** 47;49. **The direction Sine is** 112. **The unknown is** 1;56. **The known is** 132;25. **The mean [leg?] is** 510;40. **The divisor is** 2;7. **The result is**

120;40. **The square root [of that] is** 150. **[The Sine of] the zenith distance is** 30. **The shadow is** 1;8.

This is Jñānarāja's solution for the problem posed in verse 34.

Now, by means of another method the computation of the Sine of the *yojana*s in the distance [is computed] in this manner. The equinoctial shadow in Vārāṇasī is 5;40. **The [shadow] hypotenuse is** 13;17. **The Sine of the latitude is** 68;12. **The unknown is** 1;56. **[The Sine of] the amplitude is** 72;27. **The known is** 200;38. **The divisor is** 2;8. **The mean [leg?] is** 775;14. **The result is** 181;30. **The square root [of that] is** 138;0. **The Sine of the zenith distance [of the sun] is** 23;30. **Therefore, the *yojana*s in the distance are** 126.

Here an alternative solution for the solution of the problem in verse 34 is presented.

(35) **The Cosine of the degrees of the distance is divided by** 12 **and multiplied by the equinoctial shadow. From a correction by the mentioned [Sine of] the amplitude, and multiplied by the radius** R **and divided by the Sine of the distance, we get the direction Sine.**

Let s_0 be the equinoctial shadow. Let z be the sun's zenith distance and α its altitude. Then $\mathrm{Cos}(z) = \mathrm{Sin}(\alpha)$. When $\mathrm{Sin}(\alpha)$ is multiplied by $\frac{s_0}{12}$, we get the base of the Sine of the altitude. Adding or subtracting the Sine of the amplitude, we finally get the direction Sine, that is, the Sine of the degrees between the sun's current altitude circle and the prime vertical.

(36) **At night, when the sun and the moon were in Capricorn, a thief stole the best of the king's horses and quickly went** 580 ***yojana*s, and computed a shadow to the east equal to** 12**, revolving in its own direction.**

Neither the narrative nor the technical meanings are fully clear.

[The tropical longitude of] the sun is 9;0,0,0. The Sine of the declination of the sun is 65. The shadow is 12. The [shadow] hypotenuse is 17. The direction Sine is found by means of the given method [and is] 112.

This is Jñānarāja's solution sketch for the problem posed in verse 36.

(37) **According to the hemisphere, the radius is increased or decreased by the Sine of the ascensional difference and likewise the radius of the diurnal circle is increased or decreased by the earth Sine. These are the *antyā* and the *hṛti*, respectively.**

The verse defines the *antyā* and the *hṛti*, which will be used in the following.

Consider Figure 19. Let r be the radius of the diurnal circle. Then r is the length of the line GB. If we increase this length of the earth Sine, that is, the length of AG, we get the *hṛti*, which is thus the length of the line AB. In the case where BG extends beneath the horizon (that is, if it is on the other side of the point O), we need to subtract the earth Sine rather than add it.

Consider again Figure 19. The arc AG is part of the diurnal circle and thus corresponds to a time (the time it takes for the sun to traverse this arc). However, it is measured on the celestial equator, not the diurnal circle, and there is thus a segment of the celestial equator, extending beneath OC, that corresponds to the arc AG (this segment, of course, is the ascensional difference). If the radius of the celestial equator, that is, R, is increased by a length corresponding to that segment, we get the *antyā*. As before, there are also situations where the segment needs to be subtracted.

(38) **According to the hemisphere, at midday, the leg and the upright corresponding to the difference or the sum of the degrees of the local latitude and the declination [of the sun] are to be computed; they are the Sine of the zenith distance [of the sun] and [the Sine of] the altitude [of the sun], respectively.**

In this case, the Sine of the zenith distance [of the sun] multiplied by 12 and divided by [the Sine of] the altitude [of the sun] is the shadow [of the gnomon].

The shadow hypotenuse is the square root of the sum of the square of that and the square of 12.

Let ϕ be the local latitude, δ the declination of the sun, z the zenith distance of the sun, and α the altitude of the sun. The first result amounts to

$$\mathrm{Sin}(z) = \mathrm{Sin}(\phi \pm \delta) \tag{132}$$

and

$$\mathrm{Sin}(\alpha) = \mathrm{Cos}(\phi - \delta). \tag{133}$$

See *Siddhāntasundara* 2.1.27–29[a] and the commentary thereon.

The second result follows easily from similar triangles, and the third from the Pythagorean theorem.

(39) **The *antyā* at midday diminished by Versed Sine of the *asus* of the hour angle on the diurnal circle is said to be the current *antyā*.**

That multiplied by the radius of the diurnal circle and divided by the radius is the current *hṛti*.

[That] multiplied by 12 and divided by the equinoctial hypotenuse is the current [Sine of the] altitude [of the sun].

An *asu* (literally, breath) is a unit of time corresponding to the rising of 1 minute of arc of the celestial equator. All of the results given in the verse are easily verified via similar triangles.

(40) **[The first half of the verse is unclear.]**

The current [Sine of the] altitude is attained as the radius times 12 divided by the hypotenuse.

That [current Sine of the altitude] multiplied by the equinoctial hypotenuse and divided by 12 is the *hṛti*.

The first result follows from the fact that in any shadow-triangle (that is, one formed by a gnomon, its shadow, and the shadow hypotenuse), the Sine of the current altitude of the sun is to R as 12 (the height of the gnomon) is to the shadow hypotenuse.

(41) **The current *hṛti* multiplied by the radius and divided by the radius of the diurnal circle is the [current] *antyā*.**

The reverse [?] arc of the remainder of the *antyā* with half [the duration of] the day subtracted is the *asus* in the hour angle. Half the day diminished by that is the *ghaṭikās* of the elevation.

The first formula is correct. Since the current *antyā* is the arc on the celestial equator corresponding to the diurnal-circle arc represented by the current *hṛti*, the ratio $\frac{R}{r}$ can be used to find the latter from the former. Similarly, since the arc of the current *antyā* represents time since sunrise, that arc can be subtracted from the entire length of the day, giving us the "remainder of the *antyā*." If we take away half the length of the day from this remainder, we get the time since sunrise again, or the time corresponding to the current elevation of the sun.

(42) **At the intersection of the meridian and the diurnal circle it is noon. The *antyā* and the *hṛti* situated from the rising string are said to be the two hypotenuses. The difference between the rising string of Laṅkā and the local rising string is the earth Sine or the ascensional difference.**

When the radius of the diurnal circle and the radius R are diminished or increased by that according to the hemisphere, [we get] the *hṛti* and the *antyā*, [respectively].

When the sun reaches the meridian, it is noon.

(43) **The versed Sine of the hour angles of the mean [leg?] and the given *antyā* is the distance [?]. The given *antyā* is diminished by that; it is the *hṛti* made on the circle known as the earth Sine [circle]. If the upright [in a triangle] is by means of the equinoctial shadow, then what it is measured by 12? It is [the Sine of] the altitude**

in the case of the *hṛti* [?]. When the radius R by means of that is the hypotenuse, when [the Sine of] the altitude is measured by 12, what is the given hypotenuse?

The procedure is not clear.

(44) **When the oblique [?] arc of the one greater than the radius R is to be determined, subtract the radius R, [and you get] the remainder arc, respectively [?]. Increased by 5,400 it is the oblique [?] arc by means of adding the arc [corresponding to] the Sine [?]**

The procedure is not clear.

(45) **Whatever is said in the *Siddhāntaśiromaṇi*, that the six-o'clock altitude multiplied by the noon *antyā* and divided by the Sine of the ascensional difference is [the Sine of] the noon altitude, or otherwise that [the Sine of the noon altitude] is the *hṛti* multiplied by the upright produced by the equinoctial figure and divided by the corresponding hypotenuse, all that breaks down on the equinoctial day and is therefore not presented by me.**

Jñānarāja here enters a critique a verse from the *Siddhānta-śiromaṇi*.[211] The problem with the formula cited is that on the equinoctial day both the six-o'clock altitude and the ascensional difference are 0, yielding the mathematically meaningless expression $\frac{0}{0}$. However, the formula works for any other day. Jñānarāja's rejection of the formula based on division by 0 in one instance is interesting in that his predecessors (Brahmagupta, for example) attempted to define such a division.

(46) **[Thus ends the section on] the three questions for the sake of computation of time, direction, and place in the beautiful and abundant *tantra* composed by Jñānarāja, the son of Nāganātha, which is the foundation of [any] library.**

[211] *Siddhāntaśiromaṇi*, *grahagaṇitādhyāya*, *tripraśnādhikāra*, 36.

10. Chapter on Mathematical Astronomy

Section 4: *Occurrence of Eclipses*

This section formally deals with the occurrence of eclipses, but its content does not appear unified, and much of it is unclear.

(1–2) **The weekday, located from the star of the sun [?], is diminished by** 39, 30, 24, 21, 20, 20, 20, 20, 22, 26, 33, 45, 73, 200 ***palas*** **owing to the [cosmic] wind, and increased by** 400, 100, 60, 49, 44, 44, 44, 52, 72, 132, 0, 114 ***palas*** **[respectively]. At a syzygy, [the longitude of] the sun is increased by the signs and so on of what has been traversed. Its velocity is found from the day [perhaps *tithi*?]. The result consists of the minutes of arc of the [lunar] node.** 1, 2, 3, 4, 5, 5, 5, 6, **and** 7 **[?]. It [the result] is converted to degrees in the direction of the degrees of the arc of the center of the sun [?] and [applied] to the node. Otherwise, it is corrected.**

These two verses are not clear. It is not at all clear what these numbers are, or what purpose applying them to the weekday has with respect to eclipse possibilities. They possibly relate to the *nakṣatras*, but note that there are 27 *nakṣatras*, but only 26 numbers given.

(3^{a-b}) **[When]** 19;21,33,33 **[degrees] is multiplied by the current *śaka* year diminished by** 1425, **and [the result] is increased by** 4;3,32, **[we get the longitude of] the lunar node in degrees and so on [at the beginning of the current *śaka* year].**

The verse gives a method for computing the longitude of the lunar node, providing a multiplier ($19°21'33''33'''$) and an addend ($4°3'32''$) in the process (see *Siddhāntasundara* 2.1.57–64 and the commentary thereon for information on multipliers and addends). This method for determining the longitude of the lunar node is, however, peculiar. First, it gives the longitude for the beginning of *śaka* 1425, whereas the epoch given earlier (see *Siddhāntasundara* 2.1.57–64) falls about half a year later. Secondly, the multiplier and the addend that are given are not consistent with those given previously. The multiplier for the lunar node given earlier is $19°21'11''24'''$, not $19°21'33''33'''$. Furthermore, using Jñānarāja's parameters, the addend for the lunar node for the beginning of *śaka* 1425 comes out to be $1°58'46''$, not $4°3'32''$.

However, the multiplier and the addend given here are very close to what we get using the parameters given by Bhāskara II in the *Siddhāntaśiromaṇi*, namely, $19°21'33''21'''$ as the multiplier for the lunar node, and $3°16'2''$ as the addend corresponding to the beginning of *śaka* 1425.

Why a different epoch is introduced here, and why different parameters are used, is unclear.

($3^{c–d}$) **[The longitude of the lunar node] increased by 20 degrees of the sun's entry into a sign [?] [which is itself] increased by 13 degrees [or: when greater than 13 degrees]. The lunar latitude is [equal to] half of the degrees of the arc of the sun increased by the node. [The result] is multiplied by 3 and increased by 30 degrees.**

An eclipse limit could be intended here (the subject of the sentence being the shadow of the earth), but the interpretation is not clear. We are also given a formula for the lunar latitude, but the formula is defective as described.

(4) **The [true] velocity [of the sun] is divided by 5. [The result is] diminished by its own 12th part. [This] is [the diameter of] the [apparent] disk of the sun.**

[The apparent diameter of] the disk of the moon is 649 divided by the star-eaten [?].

That [apparent diameter of the disk of the moon] is multiplied by 3, and [the result] is increased by its own 10th part and diminished by the 7th part of the [true]

velocity of the sun. [This] is [the apparent diameter of] the shadow of the earth.

The [true] velocity of the moon is [equal to] the [apparent diameter of] the disk [of the moon] divided by 74.

This verse gives formulae for computing the diameters of the disks of the sun, the moon, and the shadow of the earth from the velocities of the sun and the moon. Note that as the formulae are given here, the diameters are given in *aṅgula*s rather than in minutes of arc, where 1 *aṅgula* is equal to $3'$. These formulae more properly belong to the section on lunar eclipses, but since Jñānarāja does not give them there, a brief discussion is given here.

Note that the diameters of the sun, the moon, and the shadow of the earth depend on the velocities because they depend on their distances to the earth, and these distances, in turn, depend on the velocities.

In the following, $d_{\odot}$ denotes the diameter of the disk of the sun, $d_{☾}$ the diameter of the disk of the moon, d_{Ω} the diameter of the shadow of the earth at the moon's distance, $v_{\odot}$ the true velocity of the sun, and $v_{☾}$ the true velocity of the moon.

The first formula tells us how to compute the diameter of the disk of the sun from the true velocity of the sun:

$$d_{\odot} = \frac{v_{\odot}}{5} - \frac{v_{\odot}}{5 \cdot 12} = \frac{11}{60} \cdot v_{\odot}. \tag{134}$$

It is given in both the *Śiṣyadhīvṛddhidatantra* and the *Siddhāntaśiromaṇi*,[212] though both these texts have the diameter expressed in minutes of arc rather than in *aṅgula*s.

The second formula is not clear. What exactly is meant by "star-eaten" here is unknown to me. Maybe a formula like

$$d_{☾} = \frac{v_{☾} - 715}{25} + 29, \tag{135}$$

which is given in the *Siddhāntaśiromaṇi*,[213] is intended.

[212] *Śiṣyadhīvṛddhidatantra* 5.9. *Siddhāntaśiromaṇi*, *grahagaṇitādhyāya*, *candragrahaṇādhikāra*, 8.

[213] *Siddhāntaśiromaṇi*, *grahagaṇitādhyāya*, *candragrahaṇādhikāra*, 8.

The third formula allows us to compute the diameter of the shadow of the earth at the moon's distance from the diameter of the disk of the moon and the true velocity of the sun:

$$d_{☊} = 3 \cdot d_{☾} + \frac{3}{10} \cdot d_{☾} - \frac{1}{7} \cdot v_{\odot} = \frac{33}{10} \cdot d_{☾} - \frac{1}{7} \cdot v_{\odot}. \quad (136)$$

If we use the result in the next formula, that $d_{☾} = \frac{1}{74} \cdot v_{☾}$, we get

$$d_{☊} = \frac{33}{740} \cdot v_{☾} - \frac{1}{7} \cdot v_{\odot}, \quad (137)$$

which, expressed in minutes of arc rather than in *aṅgula*s, becomes

$$d_{☊} = \frac{99}{740} \cdot v_{☾} - \frac{3}{7} \cdot v_{\odot}. \quad (138)$$

Since $\frac{99}{740} \approx \frac{2}{15}$ and $\frac{3}{7} \approx \frac{5}{12}$, this is consistent with the formula

$$d_{☊} = \frac{2}{15} \cdot v_{☾} - \frac{5}{12} \cdot v_{\odot}, \quad (139)$$

which is given in the *Siddhāntaśiromaṇi*.[214]

The final formula given in the verse,

$$v_{☾} = 74 \cdot d_{☾}, \quad (140)$$

is given in the *Siddhāntaśiromaṇi*,[215] though expressed in terms of minutes of arc rather than in *aṅgula*s there.

(5) **The square root of the difference of the square of half the sum of the diameters [of the disks] and the square of the lunar latitude is multiplied by** 60 **and divided by the difference of the velocities. [The result] is the *ghaṭikā*s of the half duration [of the eclipse]. [When] the time of the syzygy is diminished or increased by that [half duration], it is the time of first contact or the time of release [respectively]. Half of the sum of [the diameters of] the disks diminished by the lunar latitude is the obscured [part at mid-eclipse].**

---

[214] *Siddhāntaśiromaṇi*, *grahagaṇitādhyāya*, *candragrahaṇādhikāra*, 9.

[215] *Siddhāntaśiromaṇi*, *grahagaṇitādhyāya*, *candragrahaṇādhikāra*, 8.

All these results are repeated again in the next section, where they properly belong. Further discussion is found in *Siddhāntasundara* 2.5.15, 2.5.17, and $2.5.14^{c-d}$, as well as the commentaries thereon.

Note that as formulated here, that is, without specifying what disks and velocites are involved, the formulae are applicable to both lunar and solar eclipses.

(6) **The hour angle multiplied by 4 and divided by half [of the length] of the given day is applied positively or negatively to the time of conjunction according to [whether the sun is in] the western or the eastern hemisphere. [This process is to be carried out] thus again [and again], until [the time of conjunction is] correct.**

[When the longitude of] the ascendant [is] diminished by three signs[, the result is the nonagesimal, which is taken as being equal to the meridian ecliptic point]. The small Sine of the combination of the degrees of the declination of that [meridian ecliptic point] and the degrees of the local latitude is divided by 10 and increased by itself below. [The result] is the latitudinal parallax. One should correct the lunar latitude by means of it.

This verse is a somewhat confusing version of some of the results of the section on solar eclipses. Again, the material given here belongs more properly to another section.

The hour angle is the angular distance between the position of the sun and the meridian, measured as an arc on the celestial equator, which we can take to be the difference between the longitude of the sun and the longitude of the meridian ecliptic point (the intersection between the local meridian and the ecliptic). Then the first half of the verse mirrors *Siddhāntasundara* 2.6.11–$2.6.12^{b}$, except that the longitudinal parallax applied consists only of the small longitudinal parallax defined in *Siddhāntasundara* $2.6.9^{c}$–$2.6.10^{b}$ (with the longitude of the meridian ecliptic point λ_M instead of the longitude of the nonagesimal λ_V). If a denotes half the length of the arc of the diurnal circle that is above the horizon, π_λ the small longitudinal parallax, and $\lambda_\odot$

the longitude of the sun, we here have

$$\pi_\lambda = \frac{4 \cdot (\lambda_M - \lambda_\odot)}{a}, \tag{141}$$

whereas we have

$$\pi_\lambda = \frac{4 \cdot \mathrm{Sin}(\lambda_V - \lambda_\odot)}{R} \tag{142}$$

in *Siddhāntasundara* 2.6.9^{c}–2.6.10^{b}. Using only the small longitudinal parallax implies a situation where the zenith and the nonagesimal coincide. The reader is referred to the discussion there, as well as to a comparison with verses 3–5 in the *parvasambhava* section in the *Siddhāntaśiromaṇi* of Bhāskara II.

The second half of the verse tells us how to compute the latitudinal parallax π_β, and that this is to be used to correct the lunar latitude. First, we are directed to find the longitude of the nonagesimal, which is the longitude of the ascendant (the rising point of the ecliptic on the horizon) diminished by three signs. The longitude of the nonagesimal is to be taken as equal to the longitude of the meridian ecliptic point. The combination of the declination of the meridian ecliptic point and the local latitude is the zenith distance of the meridian ecliptic point (see *Siddhāntasundara* 2.6.6$^{\text{c–d}}$).

The meaning of the quantity being "increased by itself below" must be that it is increased by itself separately and then the sum is added to the original number. In other words, from x, we get $3 \cdot x$. With this interpretation, we get

$$\pi_\beta = \frac{3}{10} \cdot \mathrm{Sin}^\ell(z_M), \tag{143}$$

which is the same result as that in *Siddhāntasundara* 2.6.10$^{\text{c–d}}$, since

$$\frac{\mathrm{Sin}(z_M)}{70} = \frac{3438}{160 \cdot 70} \cdot \mathrm{Sin}^\ell(z_M) \approx \frac{3}{10} \cdot \mathrm{Sin}^\ell(z_M). \tag{144}$$

For more details, the reader is referred to the discussion of parallax in the section on solar eclipses.

(7) **Thus, the occurrence of eclipses accompanied by demonstration is given in the beautiful and abundant *tantra* composed by Jñānarāja, the son of Nāganātha, which is the foundation of [any] library.**

11. Chapter on Mathematical Astronomy

Section 5: *Lunar Eclipses*

Eclipses have always played a major role in human imagination. Most often they are seen as inauspicious omens, and thus being able to predict when they occur is significant. The computation of an eclipse is a major part of astronomy. The present section of the *Siddhāntasundara* covers how to compute a lunar eclipse.

(1–2) **The cause of eclipses of the sun and the moon is said by sages to be a darkness, the body of which is divided into a head and a tail, and which dwells at the two intersections between the ecliptic and the inclined orbit [of the moon].**

[However,] the cause of eclipses of the sun and the moon is said by people who oppose *śruti* [that is, the revealed sacred texts] and the *purāṇas* to be the moon and the shadow of the earth, respectively. They even say that in their opinion an eclipse is not [caused] by Rāhu.

According to an ancient story recorded in the *purāṇas*, solar and lunar eclipses are caused by a supernatural being called Rāhu, who attacks the sun and the moon. The Indian astronomers understood the actual reason that eclipses occur, namely, that the disk of the moon is obscured by the shadow of the earth during a lunar eclipse and that the disk of the sun is obscured by the disk of the moon during a solar eclipse. The idea that eclipses are caused by Rāhu is refuted by Lalla in the *Śiṣyadhīvṛddhidatantra*.[216] However, with the greater Indian tradition holding that Rāhu causes eclipses, the astronomers felt uneasy

[216] *Śiṣyadhīvṛddhidatantra* 20.17–27.

rejecting it altogether. In the *Brāhmasphuṭasiddhānta*, Brahmagupta, after explaining the true cause of eclipses and denying Rāhu's involvement,[217] rejects this opinion, which he ascribes to other astronomers, and continues to argue that Rāhu does indeed cause eclipses.[218] Pṛthūdakasvāmin, the commentator on the *Brāhmasphuṭasiddhānta*, suggests that Brahmagupta does this because that which is repugnant to the people should not be mentioned.[219] Here and in the following verses, Jñānarāja argues that Rāhu is indeed the cause of eclipses.

Rāhu's tail is called Ketu.[220]

(3–4) **Learned men say that although the 10-headed [Rāvaṇa] was killed by the arrow of [Rāma,] the son of Daśaratha, the 10-headed [Rāvaṇa] was [in fact] killed by [Rāma,] the son of Daśaratha; people [versed in] the *purāṇas* likewise [say this].**

[Likewise,] even if the two eclipsing bodies are the shadow of the earth and the moon, still the cause [of an eclipse] is Rāhu. Because after drawing the moon near by means of the lunar latitude, he causes the conjunction of the eclipsing body and the eclipsed body.

As is well known in Indian mythology, Rāvaṇa, the King of Laṅkā, was killed by the arrow of the god Rāma. However, while it was the arrow that killed Rāvaṇa, the real cause was Rāma, who fired the arrow.

Just as Rāma is the real cause of Rāvaṇa's death, the arrow being merely an instrument, so also is Rāhu the real cause of eclipses. It is Rāhu who draws the moon near and thus causes the conjunction of the eclipsing body and the eclipsed body.

Note that Jñānarāja here tries to reconcile ideas from the *purāṇas* with his astronomical model. See page 40 in the Introduction.

(5) **He, whose name is given as "lunar node" by great**

[217] *Brāhmasphuṭasiddhānta* 21.35–38.

[218] *Brāhmasphuṭasiddhānta* 39–48.

[219] Commentary on *Brāhmasphuṭasiddhānta* 21.43$^{\text{a–b}}$ (see Ikeyama 2002, 131).

[220] See Pingree 1990, 275.

Luminary	Diameter
The sun	6,500
The moon	480

Table 12: Diameters in *yojanas* of the disks of the sun and the moon

men, is [also] called "Rāhu" by them. When there is a rejection of him, there would always be an eclipse of the sun and the moon each month.

After stating that "Rāhu" and "lunar node" are two names for the same thing or personage, Jñānarāja argues that if Rāhu is rejected as the cause of eclipses, there would be a solar and a lunar eclipse every month. The idea, as per the previous verse, is that it is Rāhu who pulls the moon toward the ecliptic by means of the lunar latitude, and without this pull, the moon would move in a way such that there would be a solar and a lunar eclipse each month. But this is not what we experience. This is an odd statement, though, as it implies that Rāhu keeps the moon off the ecliptic.

(6) **The disk of the shadow of the earth and the disk of the moon, which can be used for obscuring, are [merely] the weapons of Rāhu during the act of an eclipse. Therefore, they are not mentioned in the *purāṇas*, the *āgamas*, and the *saṃhitās*, but they are indicated in the *veda*.**

Since the shadow of the earth and the disk of the moon are merely the weapons of Rāhu, they are not mentioned in the *purāṇas* and other sacred texts, such as the *āgamas* and the *saṃhitās*. However, according to Jñānarāja, they are alluded to in the *vedas*.

($7^{\text{a–b}}$) **The solar disk has [a diameter of]** 6,500 ***yojanas*, and the disk of the moon has [a diameter of]** 480 ***yojanas*.**

Table 12 gives the values of the diameters of the solar and

lunar disks in *yojanas* given by Jñānarāja. They are the same as those given in the *Sūryasiddhānta*.[221] Jñānarāja will give a procedure for how to determine these numbers from observation in verse 9.

To avoid confusion with the apparent diameters used in eclipse computations, the values given here will be called mean diameters in the following.

($7^{c–d}$) **The [diameter of a luminary] is multiplied by the true velocity [of the luminary] and divided by the mean velocity [of the luminary]. The apparent diameter [is found] thus.**

Neither the sun nor the moon changes its size, but as their distances to the earth vary, their apparent sizes vary as well.

Let $\bar{d}$ denote the mean diameter of the sun, $\bar{v}$ the sun's mean velocity (measured in *ghaṭikās* per civil day), and v its true velocity. Then the apparent diameter d is given by a simple proportion:

$$d = \bar{d} \cdot \frac{v}{\bar{v}}. \qquad (145)$$

The apparent diameter of the moon is computed in the same way.

(8) **The [diameter of the] disk of the moon is multiplied by the difference of the diameters of the sun and the earth and divided by [the diameter of] the disk of the sun. The diameter of the earth diminished by the result is [the diameter of the] shadow of the earth [at the moon's distance]. [When this diameter is] divided by 15, it is [measured] in minutes of arc and so on.**

As will be explained in verse 10, the light from the sun creates a shadow in the form of a cone on the other side of the earth. When the moon passes through this shadow, a lunar eclipse (partial or total, as the case may be) occurs. In order to carry out the computations for a lunar eclipse, it is necessary to know the diameter of the shadow of the earth at the distance at which the moon passes through it.

[221] See Pingree 1978a, 616.

Let $d_{\text{♁}}$, $d_{\odot}$, and $d_{\text{☾}}$ be the diameters of the earth, the sun, and the moon, respectively, in *yojanas*. In addition, let $d_{\text{☊}}$ be the diameter of the shadow of the earth at the distance of the moon. The formula given for computing this diameter is

$$d_{\text{☊}} = d_{\text{♁}} - (d_{\odot} - d_{\text{♁}}) \cdot \frac{d_{\text{☾}}}{d_{\odot}}. \tag{146}$$

The rationale behind this formula is given in verses 11–12 by Jñānarāja, at which place a discussion of it is found. See also Figure 21.

If we insert the mean diameters of the sun and the moon and use that the diameter of the earth is 1,600 *yojanas* (the radius of the earth, which we will denote by $r_{\text{♁}}$, is 800 *yojanas*[222]), we find that the mean diameter of the shadow of the earth at the moon's distance measured in *yojanas* is

$$\bar{d}_{\text{☊}} = 1600 - (6{,}500 - 1{,}600) \cdot \frac{480}{6{,}500} = 1{,}238\frac{2}{13}, \tag{147}$$

that is, a bit more than $2\frac{1}{2}$ times the mean diameter of the moon.

Rather than using *yojanas*, we want to express the diameters in minutes of arc. As there are 15 *yojanas* in a minute of arc of the lunar orbit,[223] dividing the *yojanas* of the diameters by 15 will convert them to minutes of arc.

(9) **Having learned the amount of *asus* in the rising [time] of the disks of the moon and the sun [at a time] when [each, separately, travels with its] mean velocity, the orbit [of the respective luminary] is multiplied by that and divided by the *asus* in a nychthemeron. By the application of this proportion, the *yojanas* in the respective disk [are found].**

The procedure described here is to be carried out for both the sun and the moon and will give the diameters of their disks.

[222] I have no direct reference in the *Siddhāntasundara* to this effect, but the number 800 is added to *Siddhāntasundara* 2.6.4 as numerals by most scribes, and it further follows from the circumference of the earth being given as $5,059$ *yojanas* in *Siddhāntasundara* 1.1.74 (using $\pi = \sqrt{10}$). The *Sūryasiddhānta* also gives 800 *yojanas* as the radius of the earth.

[223] See Pingree 1978a, 556.

Let us take the moon as the example in the following. At a time when the moon is moving with its mean velocity, the time it takes for its disk to rise is measured. Jñānarāja uses the time unit *asu* (literally, breath), which is equal to $\frac{1}{3{,}600}$ of a *ghaṭikā*, for measuring the rising time. Let the time of the rising of the disk be τ, and let t be the 21,600 *asus* in a nychthemeron. Further, let $\bar{d}$ be the mean diameter of the moon, and K the *yojanas* in the orbit of the moon. Then the proportion

$$\frac{\tau}{t} = \frac{\bar{d}}{K} \tag{148}$$

gives us the mean diameter.

Note that it is necessary for the moon to travel at its mean velocity to get the mean diameter. Otherwise, we will get the apparent diameter current at that time and will need to know the true velocity of the moon in order to find the mean diameter. See (145).

Pṛthūdakasvāmin gives a similar procedure for finding the mean diameter of the moon in his commentary on the *Brāhmasphuṭasiddhānta*, only he directs that one carry out the procedure every day during a lunar month and find the average rising time, which is then used to determine the mean diameter of the moon's disk.[224]

(10) **The measure at the orbit of the moon of the diameter of the cone-shaped shadow [produced] from the contact of the rays emanating from the sun and the surface of the earth and situated in space six [signs removed] from the sun is to be computed by the wise in the manner of the computation of the shadow [created] by a lamp.**

The rays of the sun hit the surface of the earth, which blocks them, thus creating a cone-shaped shadow on the other side of the earth. See Figure 21. It is clear that the center of the shadow is located on the ecliptic at exactly six signs from the sun.

Jñānarāja says that the computation of the diameter of the shadow at the moon's distance from the earth is like that of the

[224] Commentary on *Brāhmasphuṭasiddhānta* 21.11$^{\text{a–b}}$. See Ikeyama 2002, 193.

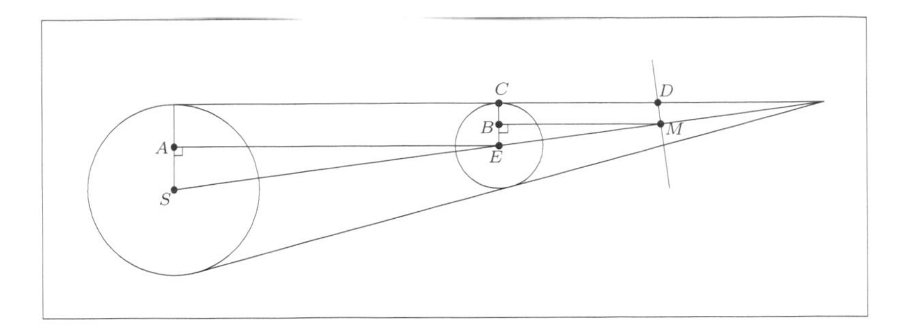

Figure 21: The radius of the shadow of the earth

computation of the shadow made by a lamp.

(11–12) **If the difference of the radii of [the disks of] the sun and the earth is the upright [corresponding to] the leg equal to the geocentric distance of the sun, then what is [the upright] when the leg is [a length inside] the shadow [of the earth], the measure of which is the mean geocentric distance of the moon? The result is at the place of the moon [that is, it is on the orbit of the moon], and is half the diameter of the earth diminished by the radius of the shadow of the earth [at the distance of the moon].**

[Having arrived at a formula] thus, here, having divided the two geocentric distances [involved in the preceding] by an [appropriate] number, [the results] are [the diameters of] the disks [of the sun and the moon], which are, respectively, the multiplier and the divisor [in the resulting formula].

The method is illustrated in Figure 21. S is the center of the sun, E the center of the earth, and M marks the distance between the center of the earth and the moon's orbit along a line through the center of the sun. The orbit of the moon is indicated through the point M.

Let $r_{\odot}$ be the radius of the sun, $r_{♁}$ the radius of the earth, $r_{☊}$ the radius of the shadow of the earth at M, $D_{\odot}$ the geocentric distance of the sun, and $D_{☾}$ the geocentric distance of the moon.

In the figure, the right-angled triangle SAE is similar to the right-angled triangle EBM, and therefore

$$\frac{|SA|}{|SE|} = \frac{|EB|}{|EM|}. \tag{149}$$

Here $|SA| = r_{\odot} - r_{♁}$, $|SE| = D_{\odot}$, and $|EM| = D_{☾}$. The radius of the shadow is $|MD|$, and if we assume that the line MD is perpendicular to the line EM, we further have that $|EB| = r_{♁} - r_{☊}$. From this we get that

$$\frac{r_{\odot} - r_{♁}}{D_{\odot}} = \frac{r_{♁} - r_{☊}}{D_{☾}}, \tag{150}$$

which, after rearranging the terms and multiplying by 2, becomes

$$d_{☊} = d_{♁} - \frac{D_{☾}}{D_{\odot}} \cdot (d_{\odot} - d_{♁}). \tag{151}$$

Finally, Jñānarāja notes that if divided by an appropriate number, $D_{\odot}$ and $D_{☾}$ become $\bar{d}_{\odot}$ and $\bar{d}_{☾}$, which is approximately correct. In fact, $D_{\odot} \approx 106 \cdot \bar{d}_{\odot}$, and $D_{☾} \approx 107\frac{1}{2} \cdot \bar{d}_{☾}$. That

$$\frac{D_{\odot}}{\bar{d}_{\odot}} \approx \frac{D_{☾}}{\bar{d}_{☾}} \tag{152}$$

is due to the fact that the disks of the sun and the moon, as seen on the sky from the earth, have roughly the same size. Using this proportion, we get

$$d_{☊} = d_{♁} - \frac{\bar{d}_{☾}}{\bar{d}_{\odot}} \cdot (d_{\odot} - d_{♁}), \tag{153}$$

which is the formula of verse 8.

Note that MD (the path of the moon through the shadow of the earth) really is an arc, not a straight line. However, given the distance of the moon from the earth, the arc can approximately be taken to be a straight line.

The Sanskrit words *śruti*, *śravaṇa*, and *karṇa*, all of which mean "hypotenuse," are used as a term for the geocentric distance of a heavenly body.

The words *dineśaśrutitulyabhāyāḥ* have been taken in the sense of "a leg equal to the geocentric distance of the sun." This length, however, is not a shadow (*bhā*).

(13) **The minutes of arc in the velocity [of the shadow of the earth or the moon] are multiplied by the *ghaṭikā*s corresponding to the end of the *tithi* and divided by 60. [The longitude of] the shadow of the earth and [the longitude of] the moon are diminished [or increased] by the result. At the end of the *tithi*, the two of them are together, having the same minutes of arc.**

The methods in the *Siddhāntasundara* give us planetary positions for midnight. In order to find the longitude at the end of the *tithi*, that is, at conjunction, we have to take into account the distance traveled by the shadow of the earth and the moon between conjunction and the time for which we have the computed positions.

Let $v_{☊}$ be the velocity of the shadow of the earth (which is the same as the velocity of the sun) and $v_{☾}$ be the velocity of the moon. Assume that there are t *ghaṭikā*s between the time for which we have the longitudes and the time of conjunction. The shadow of the earth travels $v_{☊} \cdot \frac{t}{60}$ minutes of arc and the moon $v_{☾} \cdot \frac{t}{60}$ during the time t. If the time for which we have the longitudes comes after the conjunction, we have to subtract the respective results from the longitude of the shadow of the earth and the longitude of the moon; if it comes before, we have to add them. After this operation, we have the longitudes of the shadow of the earth and the moon at the time of conjunction; the two are equal at the time of conjunction.

($14^{\text{a–b}}$) **The Sine produced from the arc [equal to the longitude] of the moon diminished by [the longitude of] the shadow of the earth is multiplied by 270 and divided by the radius. [The result] is the latitude of the moon, the direction of which is determined by the hemisphere of [the longitude of] the moon diminished by [the longitude of] the shadow of the earth.**

In the Indian astronomical system, the moon moves in an orbit that is inclined with respect to the ecliptic. It is called the inclined orbit of the moon. The inclined orbit intersects the ecliptic at two positions that are 180° from each other. The point of intersection through which the moon crosses the ecliptic moving north is called the ascending node, and the other one is called the descending node. The angle between the position of the moon and the ecliptic is called the lunar latitude, which will be denoted by β in the following.

The angle between the circle of the ecliptic and the circle of the inclined orbit of the moon is $4°30' = 270'$. This means that the greatest lunar latitude, which is attained when the moon is 90° from one of its nodes, is 270′. For other positions of the moon, the lunar latitude, measured in minutes of arc, is found by a simple proportion:

$$\beta = \mathrm{Sin}(\lambda_{☾} - \lambda_{☊}) \cdot \frac{270}{R}, \tag{154}$$

where $\lambda_{☊}$ is the longitude of the ascending node and $\lambda_{☾}$ is the longitude of the moon.

The direction of the latitude is either north or south, depending on where the moon is with respect to the ascending node. If the moon is between 0° and 180° from the ascending node (measured in the direction of motion of the moon), the direction of the latitude is north; if it is between 180° and 360° from the ascending node, the direction is south.

It is essential to know the lunar latitude when computing a lunar eclipse. The lunar latitude tells us how far the moon is from the ecliptic and thus whether we will have a total eclipse, a partial eclipse, or no eclipse at all.

($14^{\text{c–d}}$) **Half the sum of the measures [of the disks] of the eclipsed body and the eclipsing body is diminished by the lunar latitude. [The result] is the obscured [part]. This [quantity] diminished by [the diameter of the disk of the] eclipsed body is said by the wise to have the name sky-obscured [part].**

Figure 22 shows the magnitude of a lunar eclipse at mid-eclipse in three different situations. In each example, S is the center of the shadow of the earth and M the center of the moon.

In the first example, we have a partial eclipse. The length of the line CB is called the "obscured part" in the Indian tradition. We take the lunar latitude to be the line segment SM in each case. It is easy to see that

$$|CB| = (|SB| - |SM|) + |CM| = r_{☊} + r_{☾} - \beta, \tag{155}$$

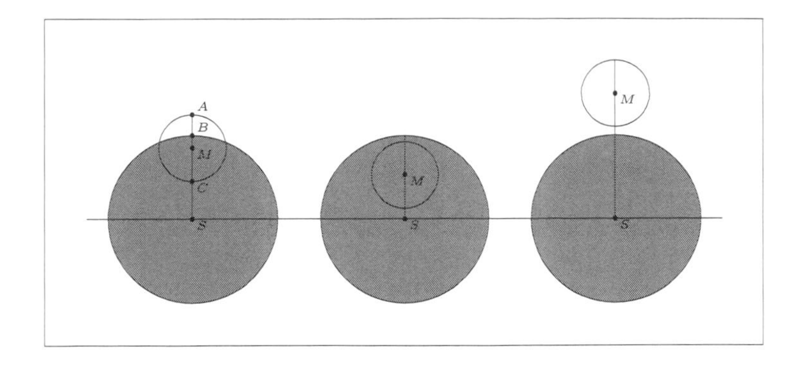

Figure 22: Examples of the lunar disk at mid-eclipse

as stated by Jñānarāja. The length of the line BA is called the "sky-obscured" part in the Indian tradition. It is equally easy to see that

$$|BA| = |CA| - |CB| = d_{\leftmoon} - (r_{\ascnode} + r_{\leftmoon} - \beta) = r_{\leftmoon} - r_{\ascnode} + \beta. \quad (156)$$

In the second example, we have a total eclipse. In this case, following the formula given by Jñānarāja, the obscured part is greater than or equal to $d_{\leftmoon}$. What this means is that the obscured part tells us how deep the moon is inside the shadow.

In the third and last example, the moon is too far from the ecliptic for there to be an eclipse.

(15–16) **The square root of the difference of the square of half of the sum [of the diameters] of the disks [of the moon and the shadow of the earth] and the lunar latitude is multiplied by 60 and divided by the [true] velocity of the moon diminished by that of the sun. The result is half of the duration of a lunar eclipse.**

Likewise, the mean half duration of totality is [found] from the difference of the halves of the measures [of the diameters of the disks].

The half duration computed from the lunar latitude current at the time of first contact or release is correct.

The eclipse begins when the disk of the moon first touches the disk of the shadow of the earth. For this to happen, the moon must be sufficiently close to the ecliptic. When the moon is first completely inside the shadow (still touching the edge of the shadow), it is said to be the beginning of totality. In other words, when the moon first touches the shadow, it is the beginning of the eclipse, and when it is first entirely covered by the shadow, it is the beginning of totality.

Let $r_{☾}$ and $r_{☊}$ be the radii of the moon and the shadow of the earth, respectively. Let $v_{☾}$ and $v_{\odot}$ be the velocities of the moon and the sun, respectively (note that the shadow of the earth moves with the same velocity as the sun). Finally, let β be the lunar latitude. Then half the duration of the eclipse is given by

$$\frac{60 \cdot \sqrt{(r_{☊} + r_{☾})^2 - \beta^2}}{v_{☾} - v_{\odot}}, \tag{157}$$

and half the duration of totality is given by

$$\frac{60 \cdot \sqrt{(r_{☊} - r_{☾})^2 - \beta^2}}{v_{☾} - v_{\odot}}. \tag{158}$$

The first of these results is derived by Jñānarāja in verses 18–19, and the commentary on these verses explains them both.

As the lunar latitude changes throughout the eclipse, we need to use the correct lunar latitude at first contact, beginning of totality, and so on in order to get the correct times.

(17) **When the time of the end of the *tithi* is diminished or increased by the [half] duration, [the result is], respectively, [the time of] first contact and [the time of] release.**

Likewise, when it is diminished or increased by the half duration of totality, [the result is, respectively,] the time of beginning of totality and the time of end of totality.

The moment of conjunction is mid-eclipse. It also occurs at the end of a *tithi*. If we assume that mid-eclipse occurs precisely at the middle of the eclipse (with respect to time), then it is clear that adding or subtracting the half duration to or from mid-eclipse yields the end and the beginning of the eclipse, and similarly for the end and the beginning of totality.

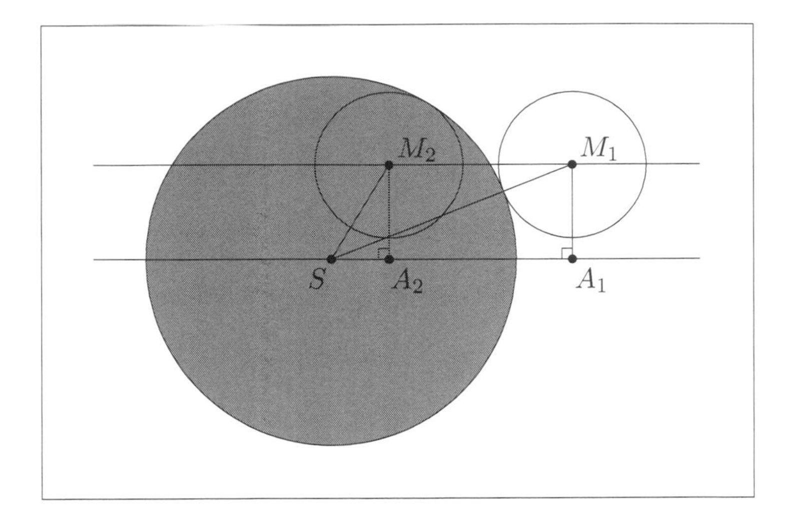

Figure 23: Half duration of the eclipse and of totality

(18–19) **The center of the shadow of the earth is on the ecliptic, and the center of the moon is at the tip of the lunar latitude on its latitude circle [that is, its inclined orbit]. At first contact, the two are at a distance equal to half of the sum of the measures [of the diameters of their disks from each other]. At mid-eclipse, they are at a distance equal to the lunar latitude [from each other]. Therefore, the hypotenuse in the [first] case is equal to half of the sum of the measures [of the diameters of the disks], the upright is the lunar latitude, and the leg is the square root of the difference of the squares [of the hypotenuse and the upright]. The half duration [of the eclipse] is that [leg], which, after [the application of] a proportion involving the velocity of the moon diminished by that of the sun, has the form of *ghaṭikā*s.**

Jñānarāja here derives one of the two formulae from verses 15–16. In Figure 23, M_1 is the position of the center of the moon at first contact and M_2 is the position at the beginning of totality. The line M_2M_1 is the orbit of the moon, and the line SA_2A_1 is the ecliptic. They are parallel, indicating that we are assuming that the lunar latitude remains constant during the eclipse.

Notice that $SM_1 = r_{\Omega} + r_{☾}$ and $SM_2 = r_{\Omega} - r_{☾}$. We will restrict ourselves to working out the formula for first contact; the procedure is analogous for the beginning of totality.

From the Pythagorean theorem we get that

$$|SA_1| = \sqrt{|SM_1|^2 - |M_1A_1|^2} = \sqrt{(r_{☾} + r_{\Omega})^2 - \beta^2}. \qquad (159)$$

Since the velocity of the moon with respect to the shadow of the earth is $v_{☾} - v_{\odot}$, it takes the moon

$$60 \cdot \frac{\sqrt{(r_{☾} + r_{\Omega})^2 - \beta^2}}{v_{☾} - v_{\odot}} \qquad (160)$$

ghaṭikās to travel the distance $|SA_1|$. The formula follows from this.

(20–21) **If the measure of obscuration at a given time is computed from [the time of] first contact or from [the time of] release], then the velocity of the moon diminished by that of the sun is multiplied by the *ghaṭikās* in the difference of the half duration and the given [time] and divided by 60. The result is the leg. The [corresponding] upright is the latitude current at the given time, and the hypotenuse is the square root of the sum of the squares of those [two quantities].**

The amount [equal to] half of the sum of the measures [of the diameters of the disks] diminished by the hypotenuse is considered to be obscured.

Let τ be the given time between the beginning and the end of the eclipse and t the half duration, both measured in *ghaṭikās*, and consider Figure 24 (where the given time corresponds to the lunar latitude being AM). Since the moon travels the distance SA in the time $t - \tau$, it is clear that

$$|SA| = \frac{(t - \tau) \cdot (v_{☾} - v_{\odot})}{60}, \qquad (161)$$

where we divide by 60 because the velocities are given with respect to civil days. This is the leg in the right-angled triangle

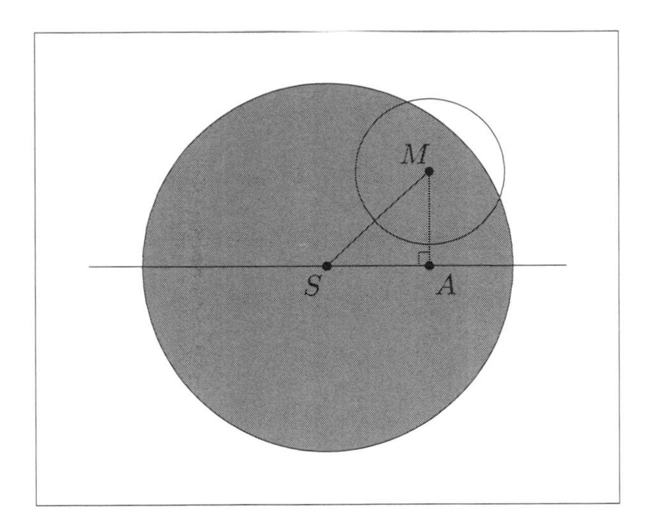

Figure 24: The obscuration at a given time

SAM, the upright of which is $AM = \beta$. The hypotenuse is then

$$|SM| = \sqrt{|SA|^2 + |AM|^2} = \sqrt{\left(\frac{(t-\tau)\cdot(v_{☾} - v_{\odot})}{60}\right)^2 + \beta^2}. \qquad (162)$$

The obscured portion of the disk of the moon at the given time is then defined as $r_{☊} + r_{☾} - h$.

As in the case of mid-eclipse, the obscured portion at a given time can be larger than the disk of the moon itself. In that case, as before, the obscured portion tells us how deep the disk is inside the shadow.

(22) **An observer has a need for first contact, mid-eclipse, and release, which are lying on the east-west [line] on the disk on its own account. Therefore, for the sake of computing directions, I will now explain the *valanas*.**

The Sanskrit term *valana* is generally translated as "deflection." The *valana* is the angle between the ecliptic and an "east-west" line on the disk of the moon. This "east-west" line is perpendicular to the great circle through the center of the moon and the north and south points on the local horizon.[225] This concept serves only divinatory purposes and has no astronomical value.[226]

(23) **One *valana* is caused by terrestrial latitude. The second is produced from the pair of *ayanas* [that is, where the bodies are with respect to the celestial equator], and the third is what is called latitudinal parallax in a solar eclipse.**

Normally Indian astronomical texts operate with only two components of *valana*,[227] the *akṣavalana*, due to the latitude of the observer's location, and the *ayanavalana*, due to the declination of the bodies.

It is peculiar that Jñānarāja here includes latitudinal parallax as a *valana* for a solar eclipse.[228]

(24–25) **With a motion along the ecliptic, the eclipsing body ... east-west. It is said that when there is no lunar latitude, it obscures the disk. Therefore, it is on its own east-west line [?].**

At mid-eclipse, the *valana*, which is the distance [?], is corrected by me. As such, it is clear and explained. At the rise of the lunar latitude from the tip of the *valana*, having given that [*valana*?], one should indicate it on the course of first contact and release.

These two verses give an explanation of the *valana*. However, the meaning is not clear. The passage marked with an ellipsis appears to be corrupted.

(26) **The Sine of the precessional [longitude of a] planet increased by three signs is multiplied by the Sine of 24 degrees and divided by the radius. The arc corresponding to that [result] is the *ayana* [*valana*], which has**

[225] See Pingree 1978a, 549.
[226] See Williams 2005, 238.
[227] See Pingree 1978a, 549 and Williams 2005, 238–239.
[228] For latitudinal parallax in a solar eclipse, see the commentary on *Siddhāntasundara* 2.6.2^{a-b}.

the [same] direction [as that] of the planet increased by three signs.

The formula given for the *ayanavalana* is

$$\mathrm{Sin}(\lambda^* + 90^\circ) \cdot \frac{\mathrm{Sin}(24^\circ)}{R}, \tag{163}$$

where λ^* is the precessional longitude of the planet. It is sometimes given using $\mathrm{Vers}(\lambda^* + 90^\circ)$ instead of $\mathrm{Sin}(\lambda^* + 90^\circ)$,[229] but Bhāskara II rejects the use of the versed Sine here.[230]

(27) **The equinoctial shadow is multiplied by the Sine of the hour angle and divided by the equinoctial hypotenuse. The *akṣavalana* is given by the degrees of the arc corresponding to the result. [Last part of the verse not clear.]**

Let d denote the hour angle (that is, the depression from the meridian), s_0 the equinoctial shadow, and h_0 the equinoctial hypotenuse. The formula given for the *akṣavalana* is

$$\mathrm{Sin}(d) \cdot \frac{s_0}{h_0}. \tag{164}$$

The last part of the verse is not clear.

(28) **When their directions are the same or different, take the sum and the difference, respectively, of the two *valanas*. The Sine of the result is divided by the radius and multiplied by half of the sum of the measures [of the diameters of the disks]. [This] is the true *valana*.**

This agrees with what Bhāskara II states in *Karaṇakutūhala* 4.16.

(29–30^{b}) **For [the sake of] understanding [of the formulae for the *valanas*], [assume that] the planet is located at one of the solstitial points and is on the meridian. In this case, the east-west [line] on the disk [of the eclipsed**

[229] See, e.g., *Śiṣyadhīvṛddhidatantra* 5.25.
[230] See *Karaṇakutūhala* 4.3.

body] along the path of the ecliptic is to be considered the "east-west" [line].

Therefore, on the terrestrial equator, there is no *valana* when the planet is located at one of the solstitial points; when [the planet] is located at the beginning of Libra or at the beginning of Aries, the distance between that [planet] and the east-west line is equal to the degrees of the greatest declination [that is, the obliquity of the ecliptic].

When the planet is located in an intermediate direction, [the *valana*] is [found] from a proportion. This is the *ayanavalana* [for a location] on the terrestrial equator.

Jñānarāja opens his explanation of the *valana* by stating that when the planet is at one of the solstitial points and on the local meridian, the "east-west" line on the disk of the eclipsed body coincides with the ecliptic. This is so because in this case the great circle through the north and south points of the local horizon and the eclipsed body and the ecliptic are perpendicular to each other.

Let us now assume that we are in a location on the terrestrial equator. If the eclipsed body is at one of the solstitial points, that is, if its tropical longitude is either 90° or 270°, then there is no *valana*. If, on the other hand, it is at one of the equinoctial points, that is, if its tropical longitude is 0° or 180°, then the *valana* is equal to the obliquity of the ecliptic. This is easy to see. In the first case, the "east-west" line is the ecliptic, and hence there is no *valana*. In the second case, the "east-west" line is the celestial equator, and hence the *valana* is equal to the angle between the ecliptic and the celestial equator, that is, the obliquity of the ecliptic.

Let the tropical longitude of the eclipsed body be λ^*, and let γ denote the *valana*. For a location on the terrestrial equator, we have that if $\lambda^* = 0°$ or $\lambda^* = 180°$, then $\gamma = \varepsilon$, and if $\lambda^* = 90°$ or $\lambda^* = 270°$, then $\gamma = 0$. From this we get that if $\text{Sin}(\lambda^* + 90°) = 0$, then $\text{Sin}(\gamma) = 0$, and if $\text{Sin}(\lambda^* + 90°) = R$, then $\text{Sin}(\gamma) = \text{Sin}(\varepsilon)$. For a tropical longitude different from the equinoctial and solstitial points, the *valana* is now found from a proportion:

$$\mathrm{Sin}(\gamma) = \mathrm{Sin}(\varepsilon) \cdot \frac{\mathrm{Sin}(\lambda^* + 90^\circ)}{R}. \tag{165}$$

On the terrestrial equator, this is the total *valana*, but elsewhere it is only one out of two components of the *valana*, called the *ayanavalana* and denoted by γ_2. In general we therefore have that

$$\mathrm{Sin}(\gamma_2) = \mathrm{Sin}(\varepsilon) \cdot \frac{\mathrm{Sin}(\lambda^* + 90^\circ)}{R}. \tag{166}$$

Note that the Sine of the *ayanavalana* equals the declination of the tropical longitude of the eclipsed body increased by 90°.

($30^{\mathrm{c-d}}$) **In a region with latitude [that is, not on the terrestrial equator], considering that it [the *valana*] is affected by the terrestrial latitude, the *akṣavalana* is applied by the ancients.**

If an observer is not on the terrestrial equator but elsewhere on earth, it is necessary to apply to the *akṣavalana* in addition to the *ayanavalana*.

(31) **The ecliptic is to be imagined as resembling the prime vertical. Whatever are the south-north [line] and one's own horizon, the termination of the horizon from that point, that *valana* has the name *ayana* in its own direction.**

(32) **On the terrestrial equator, when the planet is on the prime vertical, whatever is the east-west [line] on the disk, that is its own [east-west line]. Therefore, the *akṣavalana* is not produced [for an observer on the terrestrial equator].**

On the given horizon, the Sine of the latitude is the distance between them [the terrestrial equator and the given location]; whatever is in between by a proportion, that is said to be the *akṣavalana* by the wise. The *ayana* [*valana*] is corrected by that.

(33) **When for an observer the circle from the center of the eclipsed body, all around by means of half the sum of the apparent diameters, touches the center of the eclipsing body along a line, then there is always an eclipse.**

(34) **Having put down the *aṅgula*s of the lunar latitude in the opposite direction of first contact, then the corrected *valana* extends up to the eastern direction on the disk.**

(35) **The syzygy is computed along the circle of stars between the marks of the sun and the moon. The coming together of the disks is not there, since the moon is situated at the tip of the lunar latitude. Therefore, for the moon corrected by the visibility corrections, it is the sum of the disks.**

What is not taught by the ancients? We do not know!

(36) **The small Sine of the precessional [longitude of the] moon increased by three signs is multiplied by 13 and divided by 32. The [small] Sine of the declination is attained [as the result]. It is multiplied by the lunar latitude and divided by the [small-Sine] radius. The minutes in the result are applied positively or negatively to the application to [the longitude of] the moon depending on whether the directions of the lunar latitude and the declination are different or the same. The *tithi* at the syzygy is [determined] from the moon supplied with the visibility corrections.**

If we replace the Sines with small Sines in the formula of verse 26, we get the *ayanavalana* as

$$\mathrm{Sin}^{\ell}(\lambda^{*} + 90^{\circ}) \cdot \frac{\mathrm{Sin}^{\ell}(24^{\circ})}{R}. \tag{167}$$

Since

$$\frac{\mathrm{Sin}^{\ell}(24^{\circ})}{R} = \frac{64;20}{160} = \frac{193}{480} \tag{168}$$

and

$$\frac{13}{32} - \frac{193}{480} = \frac{1}{240}, \tag{169}$$

we see that Jñānarāja's $\frac{13}{32}$ is an approximation $\frac{\mathrm{Sin}^{\ell}(24^{\circ})}{R}$; the formula of the verse follows from this.

(37–39) **Having put down a circle using a string [measured] by the *aṅgula*s of the sum of half of the measures [of the diameters of the apparent diameters]. In the computation of the directions, the *valana* current at first contact is to be put down at the eastern [part] of the moon according to the quarters. The [*valana*] current at release is at the western [part of the moon]. In the case of a solar eclipse, it is opposite. The two latitudes [current at first contact and release] are at its [the *valana*'s] tip like Sines. The long line between the tips of the *valana*s is oblique. The mean latitude is [given] from the center according to the quarters. In whatever manner the center of the disk of the eclipsing body is at the [three] marks of the latitudes at [respectively] first contact, mid-eclipse, and release, precisely in that manner the rule regarding the directions is to be considered.**

Jñānarāja here explains how to draw the eclipse diagram. The directions are not entirely clear.

(40) **[Thus] ends the section on lunar eclipses in the beautiful and abundant *tantra* composed by Jñānarāja, the son of Nāganātha, which is the foundation of [any] library.**

12. Chapter on Mathematical Astronomy

Section 6: *Solar Eclipses*

After a careful discussion of lunar eclipses, this section turns to solar eclipses. Much of the material is the same, but solar eclipses are more complicated. During a lunar eclipse, the moon and the portion of the earth's shadow obscuring it are at the same distance from the earth, and so it is not necessary to take parallax into account. For a solar eclipse, however, it is essential to compute parallax.

(1) **Two men, one on the surface of the earth, the other at its center, do not see the sun being covered by the moon at the same time. In the case of the man at the center of the earth, the moon reaches his line of sight toward the sun precisely at the time of conjunction of the sun and the moon; it is not so in the case of the man on the surface.**

Parallax is the phenomenon that a heavenly body (in our context only the planets), when viewed from the center of the earth (we will have to postulate an imaginary "observer" there, as Jñānarāja does), is not seen at the same position with respect to the fixed stars as when it is viewed from a position on the surface of the earth. This is illustrated in Figure 25, where the sun, S, and the moon, M, are observed from the location A on the surface of the earth, as well as from the center of the earth, C. Each luminary is seen differently with respect to the fixed stars.

The parallax of a planet is the angle between the two lines formed by connecting the planet with, respectively, the center of the earth and the given location on the surface of the earth. In our example, the parallax of the sun is the angle ASC and

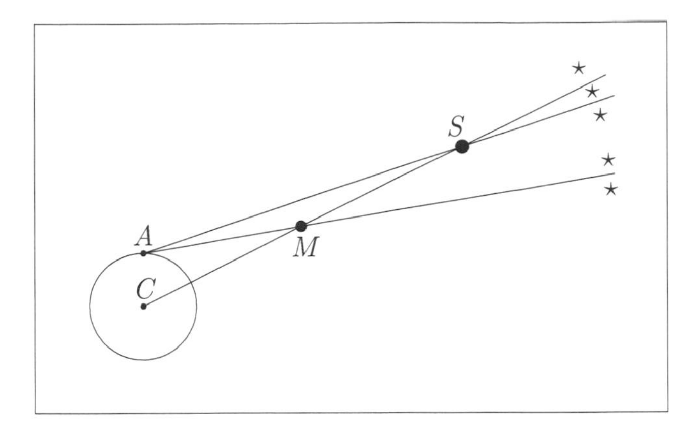

Figure 25: Parallax of the sun and the moon

the parallax of the moon is the angle AMC. As can be readily seen in the figure, the closer a planet is to the earth, the greater its parallax; the parallax of the moon is significant, while that of the sun is minor (the magnitude of the parallax also depends on the position of the planet with respect to the zenith of the observer).

When computing a lunar eclipse, it is not necessary to take parallax into account. The reason for this is that the effect of parallax is the same for the moon and the shadow of the earth, because they are seen at the same distance from the earth. However, since the sun and the moon are at different distances from the earth, the effect of parallax changes their positions not only with respect to the fixed stars, but also with respect to each other. As can be seen in Figure 25, for an observer at C, the moon is seen eclipsing the sun, whereas for an observer at A, at the same moment, no eclipse is seen. Parallax must therefore be taken into account in order to accurately compute a solar eclipse for a given locality. Note that the appearance of the sun and the moon as seen at C is not the same as that seen from most positions on the surface; if an observer is located at the point where the line CM intersects the surface of the earth, he will of course see what the "observer" at C sees.

When it comes to the role of parallax in computing a solar eclipse, what we are interested in is the combined effect of parallax on the sun and the moon: in other words, how the effect of parallax changes the positions of the two luminaries with respect to each other. We will call this the combined parallax of the sun and the moon, or simply the combined parallax. In Figure 25, the combined parallax is the angle MAS, which, seen from A, is the angular distance between the sun and the moon measured against the backdrop of the fixed stars. It is easy to see that this angle is the difference of the angles AMC and ASC:

$$\begin{aligned} \angle MAS &= 180^\circ - \angle AMS - \angle ASC \\ &= 180^\circ - (180^\circ - \angle AMC) - \angle ASC \\ &= \angle AMC - \angle ASC. \end{aligned} \tag{170}$$

So, the combined parallax of the sun and the moon is the parallax of the moon diminished by the parallax of the sun.

Since the basic theory and computations of a solar eclipse are the same as those of a lunar eclipse, this section of the *Siddhāntasundara* is devoted to an exposition of parallax. In addition to its use in computing solar eclipses, the Indian tradition also uses parallax in computing conjunction of planets.[231]

($2^{a–b}$) **The longitudinal parallax is the distance, [measured] on the ecliptic, between the lines [of sight of the two observers]. The latitudinal parallax is [measured on a great circle situated] north-south [of the ecliptic, that is, perpendicular to it].**

The effect of parallax is to make a planet appear closer to the horizon than it would be if viewed from the center of the earth. More specifically, the effect pushes a planet downward toward the horizon along a great circle through the local zenith. This can be deduced from Figure 25, where it is seen that the two lines of sight, AS and CS, both fall in the plane containing the center of the earth (C), the given location on the surface of the earth (A), the zenith corresponding to that location, and the sun (S). In other words, seen from A, the position of the sun and the position of the sun under parallax fall on the same great circle through the zenith.

Consider Figure 26. The circle $ESWN$ is the horizon with the cardinal directions marked. The line NS is the meridian,

[231] See, for example, *Siddhāntaśiromaṇi*, *Grahagaṇitādhyāya*, *Grahayutyadhikāra*, 7.

and the center of the circle, Z, is the zenith. The arc LVB is the ecliptic. The two intersections of the ecliptic and the horizon, L and B, are, respectively, the ascendant and the descendant. The intersection of the ecliptic and the meridian, M, is called the meridian ecliptic point.[232] Let P be the pole of the ecliptic, and let the intersection of the ecliptic and the great circle through Z and P be V. This point is called the nonagesimal. It is the highest point of the ecliptic above the horizon, being 90° from both the ascendant and the descendant. Further, let S' be the position of the sun on the ecliptic and S'' the sun under parallax (in other words, the angular distance $S'S''$ is the parallax of the sun). Since the effect of parallax is to shift the position of the sun downward toward the horizon along a great circle through the zenith, the points Z, S', and S'' are on the same straight line. Finally, let the arcs PS' and PS'' be parts of great circles through P. The intersection of the latter great circle and the ecliptic is called A.

The Indian astronomers separate parallax into two components. One component, called longitudinal parallax (*lambana*), is measured along the ecliptic, and the other, called latitudinal parallax (*nati*), is measured on a perpendicular to the ecliptic. The actual parallax, which is a combination of the two, will here be referred to as the total parallax.[233] In Figure 26, $S'S''$ is the total parallax, $S'A$ the longitudinal parallax, and $S''A$ the latitudinal parallax. In the following, longitudinal parallax will be denoted by π_λ, latitudinal parallax by π_β, and the total parallax by π_t.

($2^{\text{c–d}}$) **When [the longitude of] the sun is equal to [that of] the meridian ecliptic point, there is no longitudinal parallax. When the sun is at the midpoint of the prime vertical, there is neither of the two [parallaxes, that is, neither longitudinal nor latitudinal parallax].**

In this verse, Jñānarāja explains the conditions under which there will be either no longitudinal parallax or no parallax at all.

[232] This point is well known in western astrology, where it is known as midheaven.

[233] Note that there is no Sanskrit term for the total parallax, only the terms for longitudinal and latitudinal parallax.

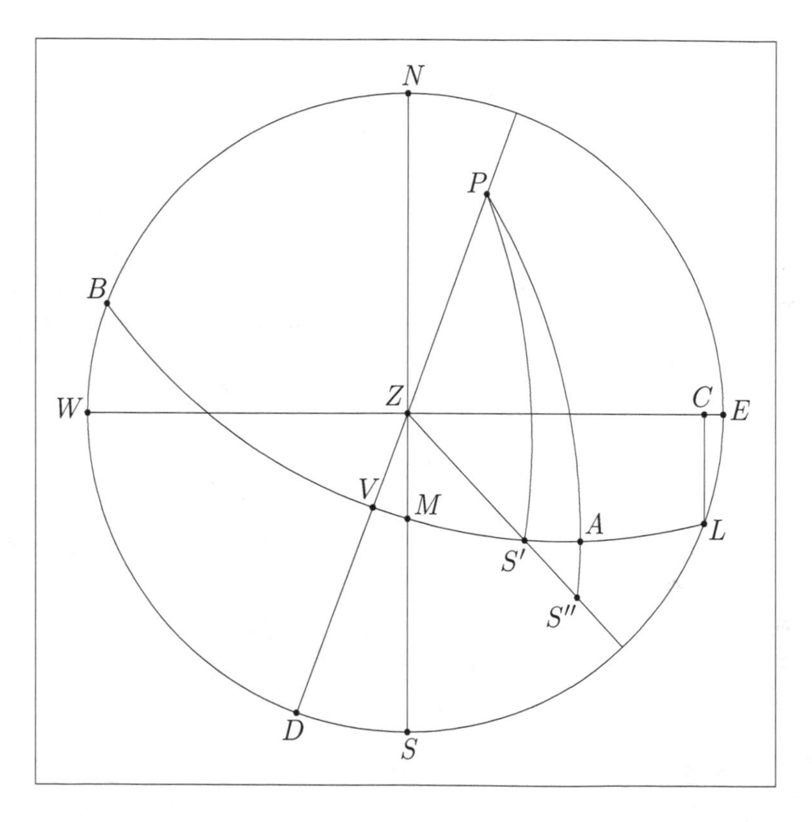

Figure 26: Projection used in computing parallax

According to him, the sun has no longitudinal parallax when it is at the meridian ecliptic point, and it has no parallax at all when it is at the midpoint of the prime vertical.

The prime vertical is the great circle through the zenith and the east and west points. In Figure 26, it is the line EZW. Its midpoint is the zenith. In other words, if the sun is at the midpoint of the prime vertical, it is at the zenith (if this is the case, the prime vertical and the ecliptic coincide). The second statement is therefore equivalent to saying that there is no parallax when the sun is at the zenith. This is a true statement. If the sun is at the zenith, the line of sight toward the sun of an observer at the given location coincides with that of an observer at the center of the earth, so that the sun will have the same position with respect to the fixed stars for both of them.

The statement is, of course, true for any planet, not just the sun. A planet located at the zenith has no parallax. Furthermore, it is easy to see that the planet's parallax increases as it gets closer to the horizon. On the horizon, a planet attains its greatest parallax.

Let us now consider the first statement and investigate under which conditions there is no longitudinal parallax. It is clear that the longitudinal parallax is nil if and only if the total parallax is perpendicular to the ecliptic. Otherwise, the total parallax would have a component along the ecliptic. Since the effect of parallax is that a planet is pushed down toward the horizon along a great circle through the zenith, this occurs only when this great circle is perpendicular to the ecliptic. This, in turn, occurs precisely when the sun is at the nonagesimal. Since the nonagesimal and the meridian ecliptic point are distinct,[234] Jñānarāja's first statement is incorrect.

It is unlikely that Jñānarāja is using a Sanskrit term normally used for the meridian ecliptic point (*madhyavilagna* in this case) to indicate the nonagesimal, for a comparison of verse $8^{\text{b–d}}$ with verse 9^{c}–10^{b} shows that he is aware of the distinction. Rather, the issue is that different texts give different accounts of when there is no longitudinal parallax. Some texts, like the

[234]It may happen that they coincide at a given point in time, but this is the exception; generally, the two points are different.

Brāhmasphuṭasiddhānta[235] and the *Siddhāntaśiromaṇi*,[236] state that it happens when the sun is at the nonagesimal, while others, like the *Sūryasiddhānta*,[237] state that it happens when the sun is at the meridian ecliptic point. Jñānarāja's statement, therefore, is not to be taken as one that he mathematically derived or observationally verified; he is merely following a tradition.

The *Sūryasiddhānta* seems to be the main source that Jñānarāja followed when writing his section on solar eclipses. He adds demonstrations, but follows the general structure of the *Sūryasiddhānta*'s chapter on solar eclipses, and gives the same formulae as in that text.

(3) **If there is no amount of degrees produced from the meridian ecliptic point, [then] the longitudinal parallax [of a planet] is measured by 4 *ghaṭikās* when it rises. That [value of 4 *ghaṭikās*] has been computed at the given time [for the rising of a planet] as well as at sunrise [in the case of the sun] by the wise sages using a variety of proportions.**

The meaning of the first line of the verse, *cen madhyalagnajanitāṃśamiter abhāvaḥ*—literally, "If there is nonexistence of the amount of degrees produced from the meridian ecliptic point"—is not wholly clear. It is reasonable, though, to take it to mean that there are no degrees separating the meridian ecliptic point and the zenith, that is, that the two points coincide. If taken in this way, the statement makes sense. When the meridian ecliptic point and the zenith coincide, the ecliptic, passing through the zenith, is perpendicular to the local horizon, and the total parallax, of the sun or another planet, therefore equals the longitudinal parallax. Since parallax is greatest on the horizon, the longitudinal parallax reaches its maximum at the time of the rising or setting of the planet. In the Indian astronomical tradition, this maximum is given as 4 *ghaṭikās*, the value given by Jñānarāja in the verse.

The verse is here taken as a general statement that for any planet, the greatest longitudinal parallax is 4 *ghaṭikās*, which,

[235] *Brāhmasphuṭasiddhānta* 5.2.

[236] *Siddhāntaśiromaṇi*, *Grahagaṇitādhyāya*, *Sūryagrahaṇādhikāra*, 2.

[237] *Sūryasiddhānta* 5.1; see the note in Burgess 1858–60, 162–164.

as we shall see below, holds true. Whether Jñānarāja intended it this way, or if he is only talking about the sun (as he did in the previous verse), is not clear. Taking it as I do, however, is the only way that I can make sense of *samaye 'bhimate*, "at a given time." If he only deals with the sun, the required time for the greatest longitudinal parallax would be sunrise, already mentioned, or sunset. Of course, what Jñānarāja is ultimately interested in here is not the longitudinal parallax of a given planet, but rather the combined longitudinal parallax of the sun and the moon.

Now, why, when parallax is an angular distance, is it here expressed in terms of a unit of time, the *ghaṭikā*? The idea is to express parallax not as the angular distance described earlier, but as the amount of time that it takes the planet in question to traverse that angular distance. In other words, what we are seeking is the amount of time that it takes the sun, or another planet, to traverse the angular distance corresponding to its greatest parallax. Note that in the Indian astronomical system, it is only the longitudinal parallax that is expressed as time, not the latitudinal parallax. The reason for this is that the longitudinal parallax is used to find the time of the apparent conjunction of the sun and the moon, whereas the latitudinal parallax is used to correct the lunar latitude. In the following, π_λ will be used for longitudinal parallax both as an angular distance and as a measure of time; the context makes it clear which is intended.

In Figure 27, the sun is on the horizon of the location A. In other words, observed from A, it has its greatest parallax. It can be seen that in this situation, the parallax of the sun, that is, the angular distance ASC, marks off a section of the sun's orbit, namely, the arc SD. Given the great distance between the earth and the sun, the length of this arc is roughly equal to the linear distance between S and B. This distance is equal to the radius of the earth. One can therefore say that the greatest parallax of a planet is the radius of the earth at the distance of the planet. In the *Siddhāntasundara*, the radius of the earth, which we will denote by $r_{♁}$, is 800 *yojana*s (see *Siddhāntasundara* 2.3.8 and commentary, including n. 222 on page 250).

It is held in the Indian tradition that "every planet travels the same absolute distance in the same interval of time."[238] Ac-

[238] See Pingree 1971, 83. The word "planet" here includes the luminaries,

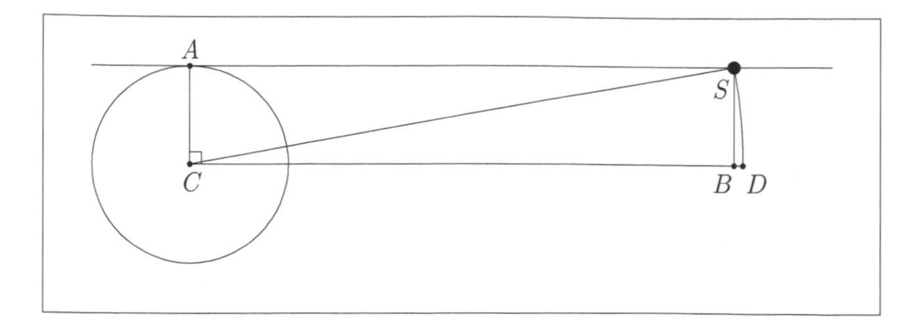

Figure 27: Greatest parallax of the sun

cording to Jñānarāja, the number of *yojanas* traversed by each planet during a *mahāyuga* is 18,712,080,864,000.[239] To find the number of *yojanas* traversed by each planet during a civil day, we divide this by the number of civil days in a *mahāyuga*,[240] which gives[241]

$$\frac{18{,}712{,}080{,}864{,}000}{1{,}577{,}917{,}828} = 11{,}858\frac{282{,}814{,}894}{394{,}479{,}457}. \tag{171}$$

If the radius of the earth is divided by this number and then expanded as a continued fraction, we get

$$\frac{800}{11{,}858\frac{282{,}814{,}894}{394{,}479{,}457}} = \frac{1}{14 + \frac{1}{1+\frac{69{,}666{,}585}{324{,}812{,}872}}} \approx \frac{1}{15}. \tag{172}$$

It thus takes a planet roughly $\frac{1}{15}$ of a civil day to traverse the angular distance corresponding to its greatest parallax. In other words, the greatest parallax of a planet is roughly a 15th part of its mean velocity. Note that using the approximation $\frac{1}{15}$ is equivalent to assuming that each planet traverses 12,000 *yojanas* per civil day.

Using this result, we find the moon's greatest parallax to be $\frac{790'35''}{15} = 52'42''$, and the sun's to be $\frac{59'8''}{15} = 3'57''$. For future reference, let us note that a more accurate computation, in which

as is our convention.

[239] See *Siddhāntasundara* 1.1.74.

[240] See *Siddhāntasundara* 2.1.15.

[241] Pṛthūdakasvāmin, in his commentary on *Brāhmasphuṭasiddhānta* 21.12, gives, using Brahmagupta's parameters, the number of *yojanas* traversed by each planet during a civil day as $11{,}858\frac{1{,}135{,}935{,}900{,}000}{1{,}577{,}915{,}450{,}000}$.

we do not use the approximation $\frac{1}{15}$, yields the moon's greatest parallax as 53′20″,[242] and the sun's as 3′59″.[243]

Now, it is clear that the greatest parallax measured in time, as explained above, is the same for all the planets, as they each travel the same distance, that is, one earth radius. We therefore need only compute it for one planet. Since the mean velocity of the moon $\overline{v}_{☾}$ is 790′35″ per civil day, it takes the mean moon 4;0 *ghaṭikā*s to traverse 52′42″ and 4;3 *ghaṭikā*s to traverse 53′20″. In both cases, it is roughly 4 *ghaṭikā*s, as stated in the verse.

In the last half of the verse, Jñānarāja says that the value of 4 *ghaṭikā*s was computed by the sages using a variety of proportions. In the next two verses, he will give a demonstration of how the greatest combined longitudinal parallax of the sun and the moon can be found, one that he presumably means to attribute to the before-mentioned sages.

(4–5) **At the time of conjunction, when the sun is on the eastern horizon for [an observer] situated at the center of the earth, the upright [of a right-angled triangle] is the radius of the earth [800 *yojana*s], the leg is the distance of the sun [from the earth] in *yojana*s, and the hypotenuse is the distance between our [location] and the sun.**

Now, [if] the leg [of a right-angled triangle similar to the one just given] is the difference of the distances of the sun and the moon [from the earth] in *yojana*s, what is the upright? It is the [combined] longitudinal parallax in *yojana*s.

[Let this combined longitudinal parallax be] multiplied by the radius and divided by the distance of the moon [from the earth]. [The result is 48;45.]

[When] 48;45 minutes of arc [are] multiplied by 60 and

[242]Since there are 15 *yojana*s per minute of arc in the moon's orbit (see verse 5.8), the moon's greatest parallax is $\frac{800}{15} = 53'20''$.

[243]It may be there, but I have not yet located a passage in the *Siddhāntasundara* that states the number of *yojana*s in the orbit of the sun. However, the value of 3′59″ for the greatest parallax of the sun can be deduced from the distance between the earth and the sun, which is given as 689,377 *yojana*s in *Siddhāntasundara* 1.1.67. The same result is arrived at if the *saurapakṣa*'s value of the number of *yojana*s in the orbit of the sun (given in Pingree 1978a, 609), that is, 4,331,500, is used.

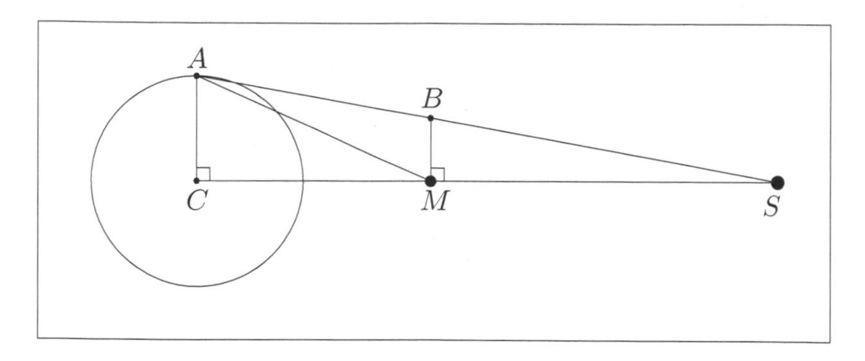

Figure 28: Figure to find the greatest combined parallax

divided by the difference of the [mean] velocities [of the sun and the moon, the result] is 4 *ghaṭikā*s. [This] is the mean longitudinal parallax at rising and setting.

In these two verses, Jñānarāja demonstrates how the value of 4 *ghaṭikā*s for the greatest combined longitudinal parallax can be found.

Consider Figure 28, which depicts the scenario described in the verses. In the figure, C is the center of the earth and S the position of the sun. Since the verse states that the sun is on the horizon for an imagined "observer" at the earth's center, we take the line CS to be his "horizon." It is further the time of conjunction, so the moon, M, is found on this line (we are not taking lunar latitude into account; during an eclipse it is not of great magnitude anyway). Further, let A be "our" location mentioned in the verse. Given the geometrical construction that Jñānarāja has in mind, the line CA has to be perpendicular on the line CS.

The triangle CAS has as its upright the radius of the earth, as its leg the geocentric distance of the sun, and as its hypotenuse the distance between the location A and the sun. This is the triangle described by Jñānarāja in the verse.

Now, let B on the line AS be chosen so that the line BM is perpendicular on the line CS. Taking into account the magnitude of the moon's distance from the earth compared to the radius of the earth, we note that $|BM|$ is approximately the greatest combined parallax of the sun and the moon (the angle

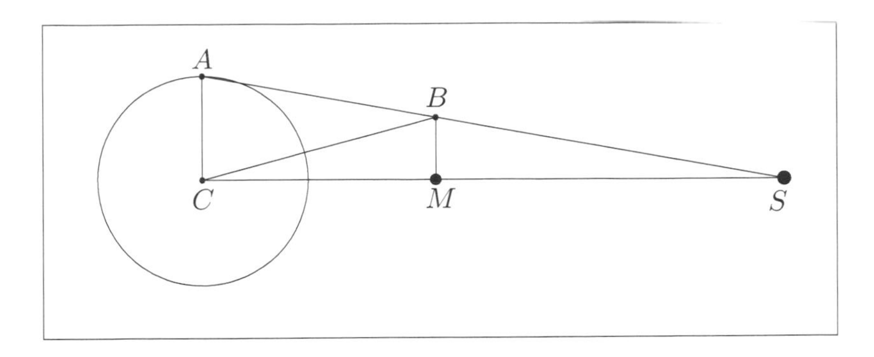

Figure 29: The greatest combined parallax as a Sine

MAS in Figure 28) measured in *yojanas* at the moon's distance from the earth. Our first step is to determine the length of BM.

Let $D_{\odot}$ denote the distance of the sun from the earth (the length CS in the figure), and $D_{☾}$ the distance of the moon from the earth (the length CM in the figure). The triangles CAS and MBS are similar, and hence

$$\begin{aligned} |BM| &= \frac{|MS| \cdot |AC|}{|CS|} \\ &= \frac{(|CS| - |CM|) \cdot |AC|}{|CS|} \\ &= \frac{(D_{\odot} - D_{☾}) \cdot r_{♁}}{D_{\odot}}. \end{aligned} \tag{173}$$

Since the radius of the earth is 800 *yojanas*, the distance of the sun from the earth is 689,377 *yojanas*,[244] and the distance of the moon from the earth is 51,566 *yojanas*,[245] we find that the length of BM equals 740;10 *yojanas*.

Now, $|BM|$ is approximately the sine of the angle BCM with respect to a circle of radius $D_{☾}$ (see Figure 29). By multiplying $|BM|$ by R and dividing it by $D_{☾}$, as Jñānarāja instructs us to do, we transform it into the Sine (that is, a sine with respect to the radius R) of that angle. The result of this operation comes out to be $\frac{740;10}{51{,}566} \cdot 3{,}438 = 49;21$, which Jñānarāja, however, gives as 48;45.

[244] See *Siddhāntasundara* 1.1.67.

[245] See *Siddhāntasundara* 1.1.66.

Since $49;21$ is a small number (compared to $R = 3{,}438$), its Sine is approximately equal to $49'21''$. In other words, the Sine of the angle BCM, which is the greatest combined parallax of the sun and the moon, is $49'21''$, or, if we use Jñānarāja's number, $48'45''$.

That Jñānarāja gets 48;45 instead of 49;21 is not surprising. We know from the notes to verse 1 that the greatest combined parallax of the sun and the moon is equal to the greatest parallax of the moon diminished by the greatest parallax of the sun. In the notes to verse 3, we found the greatest parallax of the moon to be $52'42''$ and that of the sun to be $3'57''$. The difference of these is $52'42'' - 3'57'' = 48'45''$, the result given by Jñānarāja. However, using the more accurate values computed subsequently, we get $53'20'' - 3'59'' = 49'21''$, the result we arrived at. That Jñānarāja gives $48'45''$ in the verse rather than $49'21''$ seems to indicate that he did not himself follow the computations he describes in the verses, but rather inserted an already-known result.

Finally, since the velocity of the sun and the moon with respect to each other is the difference of their respective velocities, we can convert the greatest combined parallax of the sun and the moon into *ghaṭikā*s by multiplying it by 60 and dividing it by the difference of the mean velocities of the moon and the sun, $\overline{v}_{☾}$ and $\overline{v}_{\odot}$:[246]

$$60 \cdot \frac{49'21''}{\overline{v}_{☾} - \overline{v}_{\odot}} = \frac{2{,}961}{790'35'' - 59'8''} = 4;3. \tag{174}$$

In other words, a body moving with the combined motion of the sun and the moon will take 4;3 *ghaṭikā*s to traverse the angular distance 49;21. Using Jñānarāja's value $48'45''$, we get 4;0 *ghaṭikā*s: in both cases a value close to 4 *ghaṭikā*s, as stated in the verse.

($6^{\text{a–b}}$) **[When] the Sine of 24° is multiplied by the Sine of [the longitude of] the precession-corrected ascendant and divided by the Sine of the local co-latitude, the result is the Sine of the rising amplitude of the ascendant.**

[246]The mean velocities are measured in minutes of arc per civil day; the "60" is to convert from civil days to *ghaṭikā*s.

In this and the following verses, Jñānarāja gives the formulae for computing a number of quantities that will be used to compute the longitudinal and latitudinal parallax, namely, the Sine of the rising amplitude of the ascendant, the Sine of the zenith distance of the nonagesimal, and the Sine of the altitude of the nonagesimal.

The rising amplitude of the ascendant (*udayajyā*), denoted here by η_L, is the line CL in Figure 26. The 24° mentioned is the obliquity of the ecliptic, ε. Furthermore, the declination of the ascendant is denoted by δ_L, the longitude of the ascendant by λ_L, and the longitude of the ascendant corrected for precession by λ_L^*. The latitude is denoted by ϕ and the co-latitude by $\bar{\phi}$.

The third triangle in the list of similar triangles on page 223 gives us that

$$\mathrm{Sin}(\eta_L) = \frac{R \cdot \mathrm{Sin}(\delta_L)}{\mathrm{Sin}(\bar{\phi})}. \tag{175}$$

If this is combined with the regular formula for determining the declination of the ascendant (using the usual formula for determining the declination of a point on the ecliptic given its longitude),

$$\mathrm{Sin}(\delta_L) = \frac{\mathrm{Sin}(\varepsilon) \cdot \mathrm{Sin}(\lambda_L^*)}{R}, \tag{176}$$

we get

$$\mathrm{Sin}(\eta_L) = \frac{\mathrm{Sin}(\varepsilon) \cdot \mathrm{Sin}(\lambda_L^*)}{\mathrm{Sin}(\bar{\phi})}, \tag{177}$$

which is the formula given in the verse.

In verse $13^{\mathrm{a-b}}$, Jñānarāja gives a demonstration of the formula by noting that it is derived from the two proportions (175) and (176).

($6^{\mathrm{c-d}}$) **The Sine of the zenith distance of the meridian ecliptic point is computed by means of the combination of the degrees of the [local] latitude and the degrees of the declination of the meridian ecliptic point.**

By "combination" (*saṃskṛti*) is meant that the two quantities are either added to or subtracted from each other according to whether the meridian ecliptic point is below or above the celestial equator. The zenith distance and the declination of the meridian ecliptic point (*natajyā*) are here denoted by z_M and δ_M. For the formula indicated here,

$$\mathrm{Sin}(z_M) = \mathrm{Sin}(\phi \pm \delta_M), \tag{178}$$

compare *Siddhāntasundara* 2.3.38.

(7^{a-c}) **[Let] that [Sine of the zenith distance of the meridian ecliptic point be] multiplied by the Sine of the rising amplitude of the ascendant and divided by the radius. [Take] the difference of the square of the result and the square of the Sine of the zenith distance of the meridian ecliptic point. The square root of that [difference] is the Sine of the zenith distance of the nonagesimal.**

The zenith distance of the nonagesimal (*dṛṣṭikṣepa* or *dṛkkṣepa*) is denoted by z_V.

The formula given in the verse is

$$\mathrm{Sin}(z_V) = \sqrt{(\mathrm{Sin}(z_M))^2 - \left(\frac{\mathrm{Sin}(z_M) \cdot \mathrm{Sin}(\eta_L)}{R}\right)^2}. \tag{179}$$

Jñānarāja gives a demonstration of it in verses 13^c–14^b, and a discussion of the formula is found in the commentary there.

(7^d–8^a) **The square root of the difference of that [Sine of the zenith distance of the nonagesimal] and the square of the radius is the Sine of the altitude of the nonagesimal.**

The Sanskrit phrase used by Jñānarāja to designate the Sine of the altitude of the nonagesimal is *dṛggatisañjñaśaṅkur*, "the altitude called the *dṛggati*." This usage of the term *dṛggati* is consistent with that of the *Sūryasiddhānta*,[247] but it is noteworthy that it is used differently in other texts. The *Śiṣyadhīvṛddhidatantra* and the *Vaṭeśvarasiddhānta*,[248] for example, use the term to designate the square root of the difference of the squares of the Sine of the zenith distance of the nonagesimal

[247] *Sūryasiddhānta* 5.6.

[248] *Śiṣyadhīvṛddhidatantra* 6.6. *Vaṭeśvarasiddhānta* 5.1.9.

and the Sine of the zenith distance of the sun. This is approximately $|VS'|$ in Figure 26, if the triangle ZVS' is considered planar and right-angled. It is perhaps to avoid confusion over these different usages of the term *dṛggati* that Jñānarāja specifically says that it is an altitude (*śaṅku*). Bhāskara II does not use the term *dṛggati*; he does, however, use the term *dṛṅnati* to designate what Lalla and Vaṭeśvara call *dṛggati*.

Let α_V and z_V designate the altitude and the zenith distance of the nonagesimal, respectively. The formula given in the verse,

$$\mathrm{Sin}(\alpha_V) = \sqrt{R^2 - (\mathrm{Sin}(z_V))^2}, \tag{180}$$

is straightforward. Jñānarāja gives a derivation of it in verse 14^{b}.

($8^{\mathrm{b-d}}$) **The square of half the radius divided by the Sine of [the] altitude [of the nonagesimal] is [called] the divisor. The quotient obtained from the division of the Sine of the difference between [the longitudes of] the meridian ecliptic point and the sun by the divisor is the *ghaṭikās* of the longitudinal parallax.**

In the verse, Jñānarāja merely says *nara*, "Sine of altitude," by which he must mean the Sine of the altitude of the nonagesimal. A formula for the Sine of this altitude was just given, and its appearance in the formula is consistent with the derivation of the formula in verses 14^{c}–15, as well as with the corresponding formula in the *Sūryasiddhānta*.[249] An alternative would be the Sine of the altitude of the meridian ecliptic point, which is used in the corresponding formula in the *Śiṣyadhīvṛddhidatantra*.[250]

Let $\lambda_\odot$, λ_M, and λ_V denote the longitudes of the sun, the meridian ecliptic point, and the nonagesimal, respectively. The formula of the verse,

$$\pi_\lambda = \frac{\mathrm{Sin}(\lambda_M - \lambda_\odot)}{\frac{\left(\frac{R}{2}\right)^2}{\mathrm{Sin}(\alpha_V)}} = \frac{4 \cdot \mathrm{Sin}(\lambda_M - \lambda_\odot) \cdot \mathrm{Sin}(\alpha_V)}{R^2}, \tag{181}$$

is derived by Jñānarāja from two proportions in verses 14^{c}–15, and so a discussion of it will be postponed until this point in the text.

[249] *Sūryasiddhānta* 5.7–8.

[250] *Śiṣyadhīvṛddhidatantra* 6.8.

As explained earlier, we would expect that the longitudinal parallax vanishes at the nonagesimal rather than at the meridian ecliptic point, which would require λ_V instead of λ_M in the formula (as in the formula in verse $10^{\text{a–b}}$). The formula given here is the same as that found in the *Sūryasiddhānta*.[251]

($9^{\text{a–b}}$) **This longitudinal parallax is to be applied positively or negatively to [the time of] conjunction according to whether [the longitude of] the sun is less than or greater than [that of] the meridian ecliptic point.**

Here, again, we would expect the nonagesimal where the meridian ecliptic point is mentioned. As before, Jñānarāja follows the *Sūryasiddhānta* in this respect.[252]

When, at true conjunction, the sun and the moon are to the west of the nonagesimal (this happens when the longitude of the sun is less than that of the nonagesimal), the apparent position of the moon is further to the west on the ecliptic than the apparent position of the sun. This means that the apparent conjunction has not yet occurred. Therefore, the longitudinal parallax, measured as time, is to be added to the time of conjunction.

Similarly, if the sun and the moon are to the east of the nonagesimal, the longitudinal parallax, measured as time, is to be subtracted from the time of conjunction.

To illustrate this, see Figure 26. When the sun and the moon are east of V, the moon appears farther east than the sun. That is because if the sun and the moon were together at S', and the sun is apparently displaced from S' to S'' along line ZS' as a result of parallax, then the larger parallax of the moon would displace it even further toward the horizon along the same line. This being so, its longitudinal parallax would also be larger, so it would appear east of the sun. In other words, the apparent eclipse would already have occurred, so the parallax time would be subtracted from the time of conjunction.

(9^{c}–10^{b}) **Or, [let] the Sine of the difference between the [longitudes of] the nonagesimal and the sun [be] multiplied by 4 and divided by the radius. [The result**

[251] *Sūryasiddhānta* 5.7–8.

[252] See *Sūryasiddhānta* 5.9.

is called] the small longitudinal parallax. Some people say that that [small longitudinal parallax] multiplied by the Sine of the altitude of the nonagesimal and divided by the radius is the accurate [longitudinal parallax].

The small longitudinal parallax given first,

$$\pi_\lambda = \frac{4 \cdot \mathrm{Sin}(\lambda_V - \lambda_\odot)}{R}, \tag{182}$$

corresponds to a situation where the nonagesimal and the zenith coincide. In this case, the total parallax can be found by the proportion

$$\frac{\pi_t}{\pi_0} = \frac{\mathrm{Sin}(\lambda_V - \lambda_\odot)}{R}, \tag{183}$$

where π_0 is the greatest parallax. Since in this situation the ecliptic is perpendicular to the horizon, $\pi_\lambda = \pi_t$. Using this and π_0 being 4 *ghaṭikās*, we get Jñānarāja's formula for the small longitudinal parallax.[253]

The next step is to extend the formula to the situation where the nonagesimal and the zenith do not coincide. This gives us the formula

$$\pi_\lambda = \frac{4 \cdot \mathrm{Sin}(\lambda_V - \lambda_\odot) \cdot \mathrm{Sin}(\alpha_V)}{R^2}. \tag{184}$$

With the exception that λ_M is replaced by λ_V, the formula is identical to that of verse $8^{\text{b–d}}$. For its derivation, see verses 14^{c}–15 (where Jñānarāja derives the formula of verse $8^{\text{b–d}}$) and commentary.

The formula is given in the *Brāhmasphuṭasiddhānta*[254] and the *Siddhāntaśiromaṇi*,[255] so the "some people" of the verse presumably refers to Brahmagupta, Bhāskara II, and their followers.

($10^{\text{c–d}}$) **The latitudinal parallax is [computed] from the Sine of the ecliptic zenith distance divided by 70. It is said that its direction is [the same as] that of the Sine of**

[253] The term *laghulambana*, "small longitudinal parallax," denotes the longitudinal parallax at noon in the *Vaṭeśvarasiddhānta* (5.1.21).

[254] *Brāhmasphuṭasiddhānta* 5.4.

[255] *Siddhāntaśiromaṇi*, *Grahagaṇitādhyāya*, *Sūryagrahaṇādhikāra*, 4.

the zenith distance of the meridian ecliptic point. The lunar latitude is corrected by it.

Jñānarāja now gives a formula for computing the latitudinal parallax. The same formula is found in the *Sūryasiddhānta*.[256]

To derive the formula, we consider, as is nearly correct, that the latitudinal parallax can be found by the proportion

$$\frac{\pi_0}{R} = \frac{\pi_\beta}{\mathrm{Sin}(z_V)}, \tag{185}$$

where π_0 again is the greatest parallax.

As noted under verse 3, latitudinal parallax is measured not as time, but as an angular distance. From what we found earlier, the greatest parallax of, say, the sun is $\frac{1}{15} \cdot \overline{v}_\odot$. However, we are interested in the combined latitudinal parallax of the sun and the moon, which is $\frac{1}{15} \cdot (\overline{v}_☾ - \overline{v}_\odot)$. If we insert this in formula (185), we get

$$\pi_\beta = \frac{\overline{v}_☾ - \overline{v}_\odot}{15 \cdot R} \cdot \mathrm{Sin}(z_V). \tag{186}$$

Since

$$\frac{\overline{v}_☾ - \overline{v}_\odot}{15 \cdot R} \approx \frac{1}{70}, \tag{187}$$

we get the formula of the verse. Note that according to the formula, once we know the zenith distance of the nonagesimal, the latitudinal parallax is the same at any point of the ecliptic. This is roughly correct.[257]

It is easy to see that the direction of the latitudinal parallax is the same as that of the Sine of the meridian ecliptic point. In other words, it is north when the meridian ecliptic point is north of the zenith, and south when it is south of the zenith.

The latitudinal parallax is used to correct the lunar latitude. After such a correction is made, we have the proper distance between the sun and the moon on a great circle perpendicular to the ecliptic, as seen from our location.

(11–12^{b}) **[When] the time of true conjunction is repeatedly corrected by the longitudinal parallax [until it is]**

[256] *Sūryasiddhānta* 5.11; see also *Sūryasiddhānta* 5.10.

[257] See Burgess's notes to *Sūryasiddhānta* 5.10 in Burgess 1858–60, 170–172.

constant, [the result] is the [time of] apparent [conjunction]. The lunar latitude at that [time of apparent conjunction] is corrected by the latitudinal parallax. The amount of obscuration and the half duration [of the eclipse] are [computed] from that [corrected lunar latitude]. The longitudinal parallax is to be computed [repeatedly] from the time of true conjunction increased or diminished by the half duration. Separately, by means of the accurate [longitudinal parallax], [we get the time of] first contact and the time of release, respectively.

The first step in computing a solar eclipse is to find the time of apparent conjunction, that is, the time when the sun and the moon are seen in conjunction at our location. This is done iteratively by Jñānarāja, using a method found in the *Sūryasiddhānta*[258] and the *Siddhāntaśiromaṇi*.[259]

The iterative process works as follows.[260] Let the time of true conjunction be t_0 and the longitudinal parallax at that time be π_1. When t_0 is corrected by π_1, we get a new time, say, t_1. The apparent conjunction, however, does not take place at the time t_1, because at that time, the longitudinal parallax will be different from π_1. We therefore compute the longitudinal parallax for t_1, getting, say, π_2, and correct the time of true conjunction with this new value of the longitudinal parallax. This yields, say, the time t_2. We now repeat this procedure until the times generated remain constant, say, t_n (or, equivalently, until the longitudinal parallaxes generated remain constant, say, π_n). This time is the time of the apparent conjunction.

After obtaining the time of apparent conjunction, we can compute the lunar latitude at that time. This latitude is then to be corrected by the latitudinal parallax.

Given the corrected lunar latitude, we can now, just as we did in the section on lunar eclipses, calculate the amount of obscuration of the eclipse, the half duration of the eclipse, and so on.

From the half duration we can further find the approximate

---

[258] *Sūryasiddhānta* 5.9.

[259] *Siddhāntaśiromaṇi*, *Grahagaṇitādhyāya*, *Sūryagrahaṇādhikāra*, 7.

[260] The process is described in Bhāskara II's own commentary on *Siddhāntaśiromaṇi*, *Grahagaṇitādhyāya*, *Sūryagrahaṇādhikāra*, 7.

times of first contact and release. This is done by subtracting and adding the half duration to the time of apparent conjunction, just as is the case for lunar eclipses. This will not produce the correct times, and as was the case for a lunar eclipse, we need to repeat the procedure until the times are fixed. Needless to say, Jñānarāja is rather brief in this verse, apparently expecting the reader to already be familiar with the procedure.

Once we have the accurate times of first contact, mid-eclipse (that is, the time of apparent conjunction), and release, we can proceed with the eclipse calculations as with a lunar eclipse.

Jñānarāja's text poses problems at this point. First, *sthira-vilambanasaṃskṛto* is literally "corrected by the longitudinal parallax which is constant," but the meaning has to be "until it is constant," as indicated in the translation. Secondly, the end of verse 11 is problematic, and I am not sure that my choice of reading (*pṛthak sphuṭena*) is necessarily the best one.

($12^{c–d}$) **[During a lunar eclipse,] the moon is smoke-colored when [the obscuration of its disk] is small, black when its disk is half [obscured], and tawny when the obscuration is total. [During a solar eclipse, the obscured portion of] the sun is always black.**

Schemes giving the color of the moon according to the phase of the lunar eclipse are common in Indian astronomical texts.[261] Some minor variations aside, the schemes agree with each other. Jñānarāja's scheme is unusual, though, in that it gives only three colors, whereas other texts give four.

In the case of a solar eclipse, the obscured portion of the sun's disk is always considered black in the Indian tradition.

It is interesting that Jñānarāja gives the eclipse colors in this section, and not, as is usually the case, in the section on lunar eclipses.

---

[261] See, for example, *Āryabhaṭīya*, *Gola*, 46; *Brāhmasphuṭasiddhānta* 4.19; *Śiṣyadhīvṛddhidatantra* 5.36; and *Siddhāntaśiromaṇi*, *Grahagaṇitādhyāya*, *Sūryagrahaṇādhikāra*, 36. An earlier scheme, in which the colors are assigned depending on "the altitude of the eclipsed body, its relation to the ascendant or descendant, and its magnitude" is found in the *Pañcasiddhāntikā* 6.9–10 (see Pingree 1978a, 550).

Now [three] verses [giving] demonstrations.

Demonstrations of the formulae of the section are now given.

(13^{a}) **The rising amplitude [computed] from a combination [of] two proportions that [both] involve the Sine of the declination of the ascendant at the time of conjunction [of the sun and the moon] is called the rising amplitude of the ascendant.**

The two proportions involved in computing the rising amplitude of the ascendant are (see the list of similar triangles on page 223)

$$\frac{\mathrm{Sin}(\delta_L)}{\mathrm{Sin}(\eta_L)} = \frac{\mathrm{Sin}(\bar{\phi})}{R} \tag{188}$$

and

$$\frac{\mathrm{Sin}(\delta_L)}{\mathrm{Sin}(\lambda_L^*)} = \frac{\mathrm{Sin}(\varepsilon)}{R}. \tag{189}$$

The Sine of the declination of the ascendant is found in both of them. See the commentary on verse 6^{a-b}.

(13^{b}) **The Sine of the zenith distance of the meridian ecliptic point is [found] from [the longitude of] the meridian eclictic point.**

The term *dṛgjyā* normally refers to the Sine of the zenith distance of the sun.[262] However, this cannot be the case here, because we cannot generally determine the sun's zenith distance from the longitude of the meridian ecliptic point. Considering how the text progresses in the following, the term must refer to the Sine of the zenith distance of the meridian ecliptic point.

We can compute the declination of the meridian ecliptic point from its longitude, and, in turn, its zenith distance from its declination (see verse 6^{c-d}), so the statement in the verse is clear (although, perhaps, not very profound).

[262] See, for example, *Śiṣyadhīvṛddhidatantra* 6.6; *Vaṭeśvarasiddhānta* 5.1.5; and *Siddhāntaśiromaṇi*, *Grahagaṇitādhyāya*, *Sūryagrahaṇādhikāra*, 5.

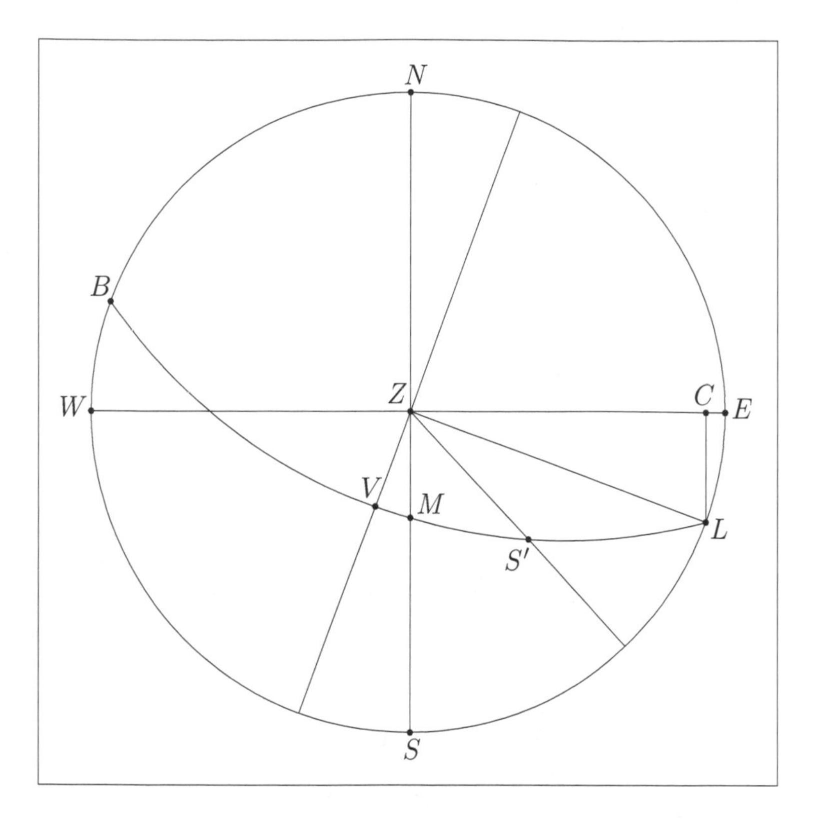

Figure 30: Sine of the zenith distance of the nonagesimal

(13^{c}–14^{a}) **[If,] when the radius [is the hypotenuse of a right-angled triangle], the leg is measured by the rising amplitude, then what is [the leg of a similar triangle, when the hypotenuse is measured] by the Sine of the zenith distance of the meridian ecliptic point? The result is a leg [corresponding to an arc that is] extending east-west on the ecliptic.**

The upright [of the triangle whose leg we just found] is said to be the square root of the difference of the square of that [leg] and the square of the Sine of the zenith distance of the meridian ecliptic point. It is the zenith distance of the nonagesimal.

Consider Figure 30, where everything is as in Figure 26. Since the lines ZL and ZV are perpendicular to each other, the angle

VZM equals the angle *LZC*. If we consider the triangle *ZMV* in the plane of the projection to be right-angled, its right angle being $\angle ZMV$ (one would expect $\angle ZVM$ to be the right angle, but that is not how Jñānarāja takes it), it is similar to triangle *ZCL*. We thus have the proportion

$$\frac{|VM|}{|ZM|} = \frac{|LC|}{|ZL|}. \tag{190}$$

This is the proportion given by Jñānarāja in the verse, since *ZL* is the radius, *LC* the rising amplitude of the ascendant, and $|VM|$ approximately the leg corresponding to the arc *VM* on the ecliptic.

From the proportion we get

$$|VM| = \frac{|ZM| \cdot |LC|}{|ZM|} = \frac{\text{Sin}(z_M) \cdot \text{Sin}(\eta_L)}{R}. \tag{191}$$

Jñānarāja next finds the upright of triangle *ZMV*, that is, the zenith distance of the nonagesimal. This is done using the Pythagorean theorem, which yields that

$$\begin{aligned} \text{Sin}(z_V) &= \text{Sin}(ZV) = |ZV| \\ &= \sqrt{|ZM|^2 - |VM|^2} \\ &= \sqrt{(\text{Sin}(z_M))^2 - \left(\frac{\text{Sin}(z_M) \cdot \text{Sin}(\eta_L)}{R}\right)^2}. \end{aligned} \tag{192}$$

We have now demonstrated the formula given in verse $7^{\text{a–d}}$.

(14^{b}) **The hypotenuse corresponding to that [upright, that is, the Sine of the zenith distance of the nonagesimal,] is the radius, and the leg corresponding to the two of them is the Sine of the altitude of the nonagesimal.**

The formula, originally given in verse 7^{d}–8^{a} and now demonstrated,

$$\text{Sin}(\alpha_V) = \sqrt{R^2 - (\text{Sin}(z_V))^2}, \tag{193}$$

is straightforward.

Here, as in verses 7^{d}–8^{a}, Jñānarāja emphasizes that the *dṛggati* is an altitude (*śaṅku*).

(14[c]–15) [If] the longitudinal parallax is 4 when the Sine of the difference of [the longitudes of] the meridian ecliptic point and the sun is the radius, what is it in the case of a given [value of the Sine of the difference of the longitudes of the meridian ecliptic point and the sun]? Then, if that [particular value of the longitudinal parallax is attained when] the Sine of the altitude of the nonagesimal is the radius, what [is it] in the case of a given [Sine of the altitude of the nonagesimal]?

In this pair of proportions, both divisors are the radius. Their product divided by 4 is furthermore [a quantity] equal to the square of the Sine of 30°. [That quantity] divided by the Sine of the altitude of the nonagesimal is called the divisor. The Sine of the difference of [the longitudes of] the ascendant and the sun divided by that [divisor] is the accurate longitudinal parallax in *ghaṭikā*s and so on.

Jñānarāja now shows how to derive the formula for the longitudinal parallax from two proportions.

If the longitudinal parallax is 4 (that is, the greatest parallax, π_0) when the Sine of the difference of the longitudes of the meridian ecliptic point and the sun is R, what is it for any given value of this Sine? Let it be π_1. Now, if this is the value of the longitudinal parallax when the Sine of the altitude of the nonagesimal is R (that is, when Z and V coincide), what is it for any given value of this Sine? Let it be π_2.

Written out as equations, the two proportions are

$$\frac{\text{Sin}(\lambda_M - \lambda_\odot)}{R} = \frac{\pi_1}{4} \tag{194}$$

and

$$\frac{\text{Sin}(\alpha_V)}{R} = \frac{\pi_2}{\pi_1}. \tag{195}$$

The first proportion corresponds to the situation where the ecliptic passes through the zenith, while the second takes the distance between the zenith and the nonagesimal into account. The second value of the longitudinal parallax, π_2, is therefore the longitudinal parallax, π_λ, that we are seeking.

Combining the two proportions will yield a formula for the longitudinal parallax. Jñānarāja combines them as follows. The divisor of each proportion is R. Dividing their product by 4 (that is, by π_0), we get $\frac{R^2}{4} = (\frac{R}{2})^2 = (\text{Sin}(30^\circ))^2$.[263] Let this quantity divided by the Sine of the altitude of the nonagesimal be called the divisor. Finally, the Sine of the difference of the longitudes of the meridian ecliptic point[264] and the sun is divided by the divisor, yielding the formula

$$\pi_\lambda = \frac{\text{Sin}(\lambda_M - \lambda_\odot)}{\frac{(\text{Sin}(30^\circ))^2}{\text{Sin}(\alpha_V)}} = \frac{4 \cdot \text{Sin}(\lambda_M - \lambda_\odot) \cdot \text{Sin}(\alpha_V)}{R^2} \quad (196)$$

as the combination of the two proportions. It is precisely the formula of verse $8^{\text{b–d}}$.

Bhāskara II, in his commentary on the *Śiṣyadhīvṛddhida-tantra*,[265] gives essentially the same demonstration. Following Lalla, he uses $\text{Sin}(\alpha_M)$ instead of $\text{Sin}(\alpha_V)$, and the two proportions are presented in the opposite order, but otherwise the demonstration proceeds exactly like Jñānarāja's.

(16) **Thus [ends] the section on solar eclipses accompanied by demonstration in the beautiful and abundant *tantra* composed by Jñānarāja, the son of Nāganātha, which is the foundation of [any] library.**

---

[263]Note that the *Sūryasiddhānta* (5.7) also expresses the formula using $\text{Sin}(30^\circ)$.

[264]Although the verse says "ascendant" (*lagna*), "meridian ecliptic point" (*madhyalagna*) is intended here.

[265]See *Śiṣyadhīvṛddhidatantra* 6.8.

13. Chapter on Mathematical Astronomy

Section 7: *Rising and Setting of Planets*

This section deals with the rising and setting of planets, including the conditions for it to happen, computing the time until the next rising or setting, and so on.

(1) **A planet with a velocity less than [that of] the sun rises [heliacally] in the east when the sun passes in front of it, and [a planet] with a greater velocity [than that of the sun] rises [heliacally] in the west when it is in front of the sun, [in both cases] according to the [appropriate] time degrees. [The planet] sets [heliacally] in the opposite manner [i.e., when slower than the sun it sets in the west and when faster than the sun it sets in the east].**

The heliacal rising of a planet in the east occurs when the planet is visible just before sunrise after not having been visible for some time due to its proximity to the sun. If the velocity of the planet is less than that of the sun, the heliacal rising in the east occurs after the sun has overtaken the planet. If the planet has a velocity greater than that of the sun, the heliacal rising in the west occurs after the planet has overtaken the sun.

Similarly, if the planet has a velocity less than that of the sun, it sets in the west when the sun catches up with it; and if it has a velocity greater than the sun, it sets in the east when it catches up with the sun.

The time degrees mentioned give the distance of the planet from the sun in order for the heliacal rising to occur; they will be given below in verse 9.

(2) **Mars, Jupiter, and Saturn, [the velocities of] which**

are less than [that of] the sun, rise [heliacally] in the east [when] their *śīghra* anomalies are 28, 14, and 17 degrees, respectively. [Their heliacal] setting in the west is said [to occur when their *śīghra* anomalies have longitudes measured] by these [values] subtracted from 360° [i.e., 332°, 346°, and 343°, respectively].

The superior planets, that is, Mars, Jupiter, and Saturn, move slower than the sun. The Indian astronomical tradition gives the values that their *śīghra* anomalies must have for the heliacal risings or settings to occur. Here values of the *śīghra* anomalies of Mars, Jupiter, and Saturn (the superior planets) are given; more values are given in the next verse.

Table 13 shows the vales of the *śīghra* anomalies given in verses 2–4.

($3–4^{b}$) **Venus and Mercury, [whose] velocities [can be] greater or less than the velocity of the sun, rise and set in the west, by means of 24, 50, 177, and 155 [degrees of their *śīghra* anomalies].**

In the east, the two [inferior planets rise and set] by 183, 205, 336, and 310 degrees.

The two inferior planets, namely, Venus and Mercury, sometimes move faster than the sun and sometimes slower.

In the west, the planet Venus has a heliacal rising when its *śīghra* anomaly is 24°, and Mercury has a heliacal rising when its anomaly is 50°; the settings occur when the *śīghra* anomalies are 177° and 155°, respectively. Similarly, in the east, Venus has a heliacal rising when its *śīghra* anomaly is 183°, and Mercury has a heliacal rising when its anomaly is 205°; the settings occur when the *śīghra* anomalies are 336° and 310°, respectively.

Table 13 shows the vales of the *śīghra* anomalies given in verses 2–4.

($4^{c–d}$) **The minutes of arc [in the current *śīghra* anomaly] increased or decreased by the given values and [then] divided by the velocity of the *śīghra* anomaly yield the days elapsed or remaining [since or until the next phenomenon].**

Planet	Rise, east	Rise, west	Set, east	Set, west
Mars	28°			332°
Mercury	205°	50°	310°	155°
Jupiter	14°			346°
Venus	183°	24°	336°	177°
Saturn	17°			343°

Table 13: *Śīghra* anomalies for heliacal risings and settings

The verse gives a simple proportion. The *śīghra* anomaly is the angular distance between the mean planet and the *śīghra* apogee. Let κ_σ be the *śīghra* anomaly, a be one of the given values (depending on what planet we are dealing with and so on), and v be the velocity of the *śīghra* anomaly. Then

$$\frac{\kappa_\sigma + a}{v} \tag{197}$$

gives the amount of time elapsed since the last heliacal rising; if we subtract a rather than adding it, we get the time remaining until the next heliacal rising.

(5) **The mean latitudes of the planets, beginning with Mars, are 90, 120, 60, 120, and 120 minutes of arc.**

[The longitudes of] the nodes [of the superior planets] are corrected by their *śīghra* equations; the nodes of Venus and Mercury are [corrected] by their *manda* equations.

By "mean latitudes" is meant the latitudes of the planets at their mean distance from the center of the earth.

Table 14 gives the values of the mean latitudes found in the verse.

(6) **[The longitude of] a planet is diminished by [the longitude of] its own node. However, in the case of Venus and Mercury its *śīghra* apogee [is diminished by the corrected node].**

[When] the arc minutes of the mean latitude [of a planet] are multiplied by the sine of [the resulting] arc

Planet	Mean latitude
Mars	$90'$
Mercury	$120'$
Jupiter	$60'$
Venus	$120'$
Saturn	$120'$

Table 14: Mean latitudes of the planets

[that is, by the sine of the angular distance between the planet and its node] and divided by the *śīghra* hypotenuse, [the results] are the true [celestial latitude of the planet], the direction of which is determined by whether the planet has passed its node.

A method for finding the true latitude of a planet at a given moment is given here. The method is fairly straightforward and mirrors what has already been given.

(7) **The sine of the declination of a [given] planet increased by 90 degrees is multiplied by the corrected latitude and divided by the radius. The resulting arc minutes are applied positively or negatively, depending on whether the direction of the declination and the latitude are the same or different[, respectively, to the corrected longitude of the planet]. [This is what is called] the *dṛggraha* [i.e., the visibility-corrected planet].**

This verse gives a formula for the visibility correction of a planet due to the deviation of the ecliptic from the orbit of the planet. The formula given by Jñānarāja differs by the one given in other astronomical texts, such as the *Śiṣyadhīvṛddhidatantra*,[266] where the versed sine is used. Indeed, below, in verse 9 of this section, Jñānarāja condemns the use of the versed sine in this particular context.

Let δ be the declination of the planet and β its latitude.

[266] *Śiṣyadhīvṛddhidatantra* 8.3.

Jñānarāja's formula for the visibility correction amounts to

$$\frac{\mathrm{Sin}(\delta + 90^\circ) \cdot \beta}{R}. \tag{198}$$

This quantity is then added or subtracted, according to whether the planet's declination has the same or a different direction, to/from the true longitude of the planet.

(8) **When the latitude is to the south, the result of the latitude multiplied by the equinoctial noon shadow and divided by 12 is [the visibility correction]; it is [applied] negatively on the eastern horizon and positively on the western; when the [latitude] is to the north, [it is applied] positively and negatively [instead].**

Thus, [the longitude of] the planet corrected for the local latitude and for the deviation of the ecliptic is the rising and setting point[, respectively,] of the ecliptic.

The formula gives the visibility correction due to local latitude. When applied in the manner described in the previous verse, the result is the point of the ecliptic that rises or sets with the planet. The formula given here agrees with the one given in the *Śiṣyadhīvṛddhidatantra*.[267]

The two corrections thus given in verses 7 and 8 give, respectively, the rising and the setting point of the ecliptic when applied to the true longitude of the planet.

($9^{\text{a–b}}$) **Those who say that the visibility correction is [computed] by means of the versed sine are deviating from the opinion of Brahmā, Sūrya, and Candra.**

As has already been noted in the commentary following verse 7, the formula given by Jñānarāja differs from that found elsewhere. In particular, the versed sine is sometimes used. In this verse, Jñānarāja condemns the use of the versed sine on the grounds that it differs from what has been taught by the deities Brahmā, Sūrya, and Candra, that is, that the formulae given in the *Brahmasiddhānta*, the *Sūryasiddhānta*, and the *Somasiddhānta* do not make use of the versed sine.

[267] *Śiṣyadhīvṛddhidatantra* 8.3.

Planet	Time degrees
moon	12°
Mars	17°
Mercury	13°
Jupiter	11°
Venus	9°
Saturn	15°

Table 15: Time degrees of the planets

(9^{c-d}) **The time degrees beginning with the moon are, in order,** 12, 17, 13, 11, 9, **and** 15.

The time degrees given are 12° for the moon, 17° for Mars, 13° for Mercury, 11° for Jupiter, 9° for Venus, and 15° for Saturn. What this means is not specified by Jñānarāja, but the heliacal rising of each of the planets occurs when it is separated from the sun by the specified amount of time degrees.

Table 15 gives the time degrees for each of the planets mentioned in the verse.

(10) **When the day is known from the degrees of the greatest equation [of a planet, that is, when the planet has its maximum velocity] [?], the visibility-corrected planet is to be used at the time of sunrise or sunset. The greater of [the longitude of] the sun and [the longitude of] the visibility-corrected planet is the rising point of the ecliptic. The sun is imagined to be west or east [of the planet] by *ghaṭikās* in the distance [between them, that is, the time degrees].**

The verse is not clear.

(11–12) **On the western [horizon], [the true longitude of the sun and that of the visibility-corrected planet] is multiplied by six and increased by six [zodiacal signs]. If the [time] degrees [of a planet] are greater than what has been stated [that is, the time degrees given], the**

[heliacal] rising has taken place; otherwise, it will occur in the future. It is the opposite for the [heliacal] setting.

The arc minutes produced from the difference between the stated and the lapsed time degrees are multiplied by 300 and divided by the rising of the sun [?] and its own seventh part. On the western [horizon], the result is divided by the difference of the velocities [of the sun and the visibility-corrected planet] through these days [?]. In the case of retrograde [motion], the division is by the sum of the velocities.

The [heliacal] rising or setting of the planet is to be understood according to the sun being on the western or eastern [horizon].

Neither of the formulae of these two verses is clear. While adding the six zodiacal signs on the western horizon makes sense, multiplying by six does not (the order of the two operations also is not clear). Similarly, the second formula presumably aims at finding the time lapsed since or remaining until a heliacal rising, though it is not clear what one is supposed to divide by in the division.

(13) **On the eastern [horizon], when [the true longitude of] the visibility-corrected planet is greater or smaller than [that of] the sun, the planet [is called] western-visibility-corrected.**

[Whether] the beginning of the day has lapsed or is to come is to be computed by means of the measure of the degrees [corresponding to] the stated and the given time, in this case for the sake of [the time of] setting or rising.

(14) **When the position of a planet is on the circle of the stars at the tip of the degrees of the declination of the mean [planet] [?], then its disk is located on the latitude circle [?] at the tip of its corrected declination [?]. The setting and rising of the disk and that [?] do not occur at one single time [?]. The two visibility corrections are to be applied on account of the latitude [of the planet].**

For the terrestrial latitude there is no rising in the case of the planet [?].

The verse is unclear.

(15) **The difference between the degrees of the time of rising or of setting of the visibility-corrected planet and the sun is the time degrees.**

When [the planets undergo] retrograde [motion], [the time degrees] are increased by one due to the disk [of the planet] being wide.

Another method that is complete and easy has never been stated in this regard.

The first part of the verse, where the rising of the planet refers to its heliacal rising, is a clear definition of the time degrees. The second part, however, is not clear since there is no correlation between the size of a planet's disk and whether or not it is undergoing retrograde motion.

(16) **[When] the position of a planet moves on the path [of the sun, that is, the ecliptic,] while on the horizon of Laṅkā, then the two poles of the ecliptic are located to the south and the north on the horizon.**

When its latitude faces [one of the ecliptic poles], then the disk position is said to be simultaneous [?].

The last part of the verse is unclear.

(17) **Therefore, in this case, for the planet is on the ecliptic here, there is no visibility correction designated by [the word] *ayana* [i.e., due to the deviation of the ecliptic], and there is no *valana* designated by that [same name] on the terrestrial equator, when its position is at the beginning of Aries or at the beginning of Libra. The two poles of the ecliptic are raised by the sine of 24 degrees [that is, the sine of the obliquity of the ecliptic].**

Regarding there being no *ayana valana* on the terrestrial equator, see *Siddhāntasundara* 2.5.29–30.

(18) **When the disk [of a planet] is located at the tip of its latitude while facing the [pole of the ecliptic], the following position is on the horizon; in this case, the visibility correction is explained for [the region] without latitude [that is, the terrestrial equator]. In a region with latitude [that is, in a place other than on the terrestrial equator], the correction due to the deviation of the ecliptic varies due to the variety of horizons.**

The meaning is not entirely clear.

(19) **[Thus] are the rising and settings of the planets explained and the visibility correction taught with demonstration in the beautiful *tantra* composed by Jñānarāja, the son of Nāganātha.**

14. Chapter on Mathematical Astronomy

Section 8: *Shadows of Stars, Constellations, Polestars, and So On*

This section of the *Siddhāntasundara* deals with a broad content. Among other things, longitudes and latitudes of constellations and stars are given.

As has been noted in the Introduction (see page 26), verses 14–23 of this section occur again in some manuscripts as a separate section between the present sections 10 and 11. In this translation, the verses are kept in section 8.

(1) **The shadow from a planet or a star [can], like the shadow of the sun, [be found] by means of the sines of the corrected declination, ascensional difference, and so on. In that case, the lapsed *ghaṭikā*s are [measured] from the rising of the planet and the remaining *ghaṭikā*s [i.e., those left until] the setting [of the planet].**

The shadow thrown by a gnomon due to the light of the sun has already been utilized in the *Siddhāntasundara*. Jñānarāja here tells his readers that one can similarly make use of the light from a planet or a star. Theoretically this is true, of course, but the practical use is rather doubtful.

(2) **When the ascensional difference is computed by means of the mean or the true declination [of the given planet], their [?] difference or sum [is found] according to whether they have the same or different directions.**

The rising point [of the planet is found] from [the resulting] *prāṇa*s and the planet corrected for visibility and for precession, in regular order or inverse order according to whether [the planet's] latitude is to the south

or to the north.

The rising and setting points of the planet [are found] by means of the *ghaṭikā*s [corresponding to] the distance between the [planet] and the sun.

The ideas expressed in the verse are not entirely clear. For example, the first part appears to refer to the difference between the mean and the true declination, though this makes little sense.

A *prāṇa* is a unit of time given by $\frac{1}{360}$ of a *ghaṭikā*.

(3) **A planet rises, during the day or at night, by means of the *vighaṭikā*s produced from the difference between six zodiacal signs and [the longitude of] the sun. The lapsed *ghaṭikā*s of the planet increased by the [result] at night or by day are lapsed.**

(4–9) **The polar longitudes of the *nakṣatra*s are in order: 8, 20, 38, 50, 63, 67, 93, 106, 108, 127, 147, 155, 180, 183, 199, 212, 224, 228 [or?] 229, 241, 254, 260, 278, 290, 320, 326, 337, and 0.**

When the time [degrees] are diminished by the degrees of precession they are said to be the polar longitudes.

The polar latitudes [of the *nakṣatra*s] are: For [the one shaped like] the head of a horse, [that is, Aśvinī,] it is 10; for the one shaped like a triangle, [that is, Bharaṇī,] it is 12; for the one shaped like a blade, [that is, Kṛttikā,] it is 2;30; for the one shaped like a cart, [that is, Rohiṇī,] as well as for the one shaped like [the head of] a deer, [that is Mṛgaśiras,] it is 10; for the star of Śiva, [that is, Ārdrā,] it is 11; for the one shaped like an uneven quadrilateral, [that is, Punarvasu,] it is 6; for the star Puṣya, which is lusterless, it is 0; for the one shaped like a serpent, [that is, Āśleṣa,] it is 7; for the one of uneven lines, [that is, Maghā,] it is 0; for the constant pair, [that is, Pūrvaphālgunī and Uttaraphālgunī,] it is 12 and 11; for the one shaped like a hand, [that is, Hasta,] it is 11; for the one shaped like a pearl, [that is, Citrā,] it is 1;45; for the one shaped like a new leaf, [that is, Svāti,]

it is 37; for [the one shaped like] an arched gateway, [that is, Viśākhā,] it is 1;20; for the one shaped like an oblation, [that is, Anurādhā,] it is 1;45; for [the one shaped like] a triangle, [that is, Jyeṣṭhā,] it is 3;30; for the pair, [that is, Mūla,] it is 8;30; for [the one shaped like] a bed, [that is, Pūrvāṣāḍā,] it is 5;20; for [the one shaped like] an elephant's tusk, [that is, Uttarāṣāḍā,] it is 5; for Abhijit, it is 62; for the one shaped like an arrow, [that is, Śravaṇa,] it is 30; for [the one shaped like] an earring, [that is, Dhaniṣṭhā,] it is 36; for [the *nakṣatra* of] 100 stars, [that is, Śatabhiṣaj,] it is 0;20; for the pair of dead and new, [that is, Pūrvabhādrapadā and Uttarabhādrapadā,] it is 24 and 26; and for the one shaped like a *mṛdaṅga* drum, [that is, Revatī,] it is 0.

The six verses give the polar longitudes and polar latitudes of the 27 *nakṣatras*, or constellations along the ecliptic. Each *nakṣatra* is, of course, a collection of a number of stars, and the coordinates given are those of the junction stars, prominent stars, one for each *nakṣatra*, with which each *nakṣatra* is identified.

The data are summarized in Table 16. Note that Jñānarāja does not specify whether the polar latitude is northern or southern; the table includes this information as given in the *Sūryasiddhānta*.[268]

The reading of the polar longitude of Jyeṣṭhā is problematic, as the text literally gives the digits of the degrees as 2298. The intended meaning could be that it is one of 228° and 229°, both of which are given in other astronomical texts; in other words, an extra and incorrect '2' has been inserted at the beginning of the number.

No polar longitude is given for the *nakṣatra* Abhijit, but the *Sūryasiddhānta*[269] gives its polar longitude and latitude as 266°40′ and 60°. Abhijit is not a regular *nakṣatra*, but rather occupies part of Uttarāṣāḍā and Śravaṇa, and it is sometimes used as the 28th *nakṣatra*.

A *mṛdaṅga* is a particular kind of clay drum, also known as a khol, which is used in northern and eastern India.

---

[268] *Sūryasiddhānta* 8.2–9.

[269] *Sūryasiddhānta* 8.2–9.

Nakṣatra	Polar longitude	Polar latitude
Aśvinī	8°	10° N
Bharaṇī	20°	12° N
Kṛttikā	38°	2°30′ N
Rohiṇī	50°	10° S
Mṛgaśiras	63°	10° N
Ārdrā	67°	11° S
Punarvasu	93°	6° N
Puṣya	106°	0°
Āśleṣa	108°	7° S
Maghā	127°	0°
Pūrvaphālgunī	147°	12° N
Uttaraphālgunī	155°	11° N
Hasta	180°	11° S
Citrā	183°	1°45′ S
Svāti	199°	37° N
Viśākhā	212°	1°20′ S
Anurādhā	224°	1°45′ S
Jyeṣṭhā	228° or 229°	3°30′ S
Mūla	241°	8°30′ S
Pūrvāṣāḍā	254°	5°20′ S
Uttarāṣāḍā	260°	5° S
Śravaṇa	278°	30° N
Dhaniṣṭhā	290°	36° N
Śatabhiṣaj	320°	0°20′ S
Pūrvabhādrapadā	326°	24° N
Uttarabhādrapadā	337°	26° N
Revatī	0°	0°

Table 16: Polar coordinates of the *nakṣatras*

Star	Polar longitude	Polar latitude
Sirius	86°	40° S
Canopus	87°	77° S

Table 17: Polar coordinates of Sirius and Canopus

(10) **[The junction star of] Āśleṣa is 3 [degrees] from Brahmā's star, 2 [degrees?] from [the junction star of] Hasta, and six regents increased by 2 [?] from Varuṇa [?]. [Its] greatest [distance] from the ecliptic, to the south or to the north, is due to the degrees in its latitude.**

The import of the verse is unclear, though it intends to give the polar coordinates of the star Āśleṣa. Varuṇa is the name of a god, the ruler of the western quarter, so the direction of west could be intended.

(11) **[The polar longitude of] Lubdhaka [that is, the star Sirius] is 86 [degrees], and its southern polar latitude is 40 degrees. [The polar longitude of] Agastya [that is, the star Canopus] is 87 [degrees], and its southern polar latitude is 77 [degrees].**

Here the polar longitudes and latitudes of two more stars are given, namely, those of Sirius and Canopus. The data are given in Table 17.

(12–13) **Now two verses giving the demonstration: After observing the given junction star at noon by means of a hollow reed positioned at the tip of a post and using a string as the path of the reed, the hypotenuse [is found]. The post is the gnomon. The difference between them is the shadow. [When] it [that is, the shadow] is multiplied by the radius R and divided by the shadow hypotenuse, the arc corresponding to the result is the degrees of the meridian-zenith distance; as explained in the discussion of planets, the polar longitude of the star is the celestial latitude.**

The intended meaning of observing the planet "at noon" is most likely on the noon of that particular planet's day, i.e., when the planet is on the meridian. The right triangle arrived at, as well as its use, has already been discussed in connection with planets.

(14–16) **[The following is] said in the *Brahmasiddhānta*: "[The star] Vasiṣṭha is 10 degrees west of [the star] Marīci. [The star] Aṅgiras is 7 [degrees] west of the [star Vasiṣṭha]. [The star] Atri is 8 [degrees] west [of Aṅgiras]. [The star] Pulastya is 3 [degrees] west [of Atri]. [The star] Pulaha is 10 [degrees west of Pulastya]. [The star] Kratu is 3 [degrees from Pulaha]. At the beginning of the *yuga*, Kratu was 5 degrees north at the beginning of Viṣṇu's *nakṣatra* [that is, Śravaṇa].**

The [northern distances from the ecliptic] of the [seven] sages are, in order, 55, 51, 50, 56, 57, 60, and 60 [degrees]. Their motion is 8 arc minutes [per year] eastward."

[With] their exceedingly small north-south motion [the seven sages] complete a revolution in 2,700 years.

The seven stars given in the verse are the ones known as the *saptarṣis*, the seven sages. As the names tell us, they are named after and represent seven sages of the Indian tradition: Marīci, Vasiṣṭha, Aṅgiras, Atri, Pulastya, Pulaha, and Kratu. Besides their positions, we are told that they make a slight eastward motion, 8′ per year, meaning that they complete a full revolution in 2,700 years.

An English translation of the passage from the *Brahmasiddhānta* cited by Jñānarāja can be found in Colebrooke 1977, 2.358.

For an overview of the idea in the Indian tradition of the seven sages having an independent motion, see Dhavale 1969.

(17) **At the beginning of the *yuga*, an eastward motion on the circle of stars [commences, that is, the zodiacal signs change their position with respect to the celestial equator] over 27 degrees, taking 1,800 years. [Then], the**

journey of its position becomes western owing to retrogradation, [traversing again] 27 degrees.

The verse explains the rate of trepidation (see the notes to *Siddhāntasundara* 2.4.44 for more information).

(18–19) **At the end of the *yuga*, [there have] thus [been] 600 revolutions in 7,200 lapsed years. What 108 by the remaining years? [?] It is multiplied by 27; the degrees of precession, known to be an additive or positive [contribution] according to whether [the motion is] to the west or to the east, are [found] in the remainder; [this is] for the successful [computation] of the declination, and so on. The degrees of precession are [found] from the difference of [the longitude of] the sun computed from [the time of] sunrise, the amplitude, and the declination.**

The verses clearly aim to compute the amount of precession, but the numbers are unclear.

(20–21) **[The stars] Hutabhuj and Brahmahṛdaya are in the 22nd degree of Taurus with latitudes 8 and 30 degrees [respectively]. It is said by the ancient sages that [the star] Prajāpati is 5 degrees east of [the star] Brahmahṛdaya having a northern latitude of 38 degrees. Apāṃvatsa is 5 [degrees] north of Citrā, and Āpas is 6 degrees north of Citrā.**

Positions of a number of additional stars are given here.

(22) **The star at the head of the fish, which is the polestar, is raised according to the degrees of the local terrestrial latitude. The tail of the fish revolves around it along with the stars [of the Polar Fish constellation] just like a young calf tied to a post.**

The polestar fish is a constellation that revolves around the fixed polestar, which forms its head. The words *padeśe nayanapade* of the verse are unclear and not translated in the above.

(23) [Thus] is the shadow of stars and the polar longitudes of stars explained in the beautiful and abundant *tantra* composed by Jñānarāja, the son of Nāganātha, which is the foundation of [any] library.

15. Chapter on Mathematical Astronomy

Section 9: *Elevation of the Moon's Horns*

With the exception of the four quadratures of the moon (that is, when the sun and the moon are in conjunction, and the moon is invisible; at full moon; and at the two points where precisely half of the moon's disk is illumined by the sun), the moon displays "horns." When the illumined portion of the disk is smaller than the dark portion, the horns are said to be "bright"; if it is opposite, they are said to be "dark." This section deals with computing the extent of the moon's horns.

(1) **The sun, possessed of strong heat, is the king in the orb of heaven. The moon, full of the digits caused by the reflection of [the sun's rays], is the minister. Possessed of the mark of its rays, [the moon,] displaying his own light, goes behind and in front of the sun to destroy the darkness, his enemy.**

The verse poetically describes the sun as the king of heaven and the moon as his minister, who aids in dispelling darkness. It is possible that the word "darkness" (*tamas*) refers to Rāhu, the demon that causes eclipses according to Indian mythology.

(2) **[The man] bearing the name Jñāna[rāja] asks the excellent mathematicians: "If the sun due to its reflection in the orb of the moon, which is made of water, is lackluster by day, how is it bright at night? What is the same as the disk of the sun? Just as [only] the half of the orb [of the moon] that is made luminous by the light of the sun, not the whole [orb], is seen in a pool [of water] at the end of the *pakṣa*."**

The text does not make complete sense here, and Jñānarāja's question to the learned mathematicians is unclear.

(3) **The orb of the moon, which is full of lotus balls, shines like pieces of lotus flowers. The half of the moon orb that is facing the sun is shining owing to the spread of the sun's rays.**

The orb of the moon is here said to made up of lotuses. The notion that only half of the moon's orb is illumined by the rays of the sun is well known.

(4) **The gods drink the orb of nectar during the days following the first day of the dark *pakṣa*, [that is, when the moon is waning]. That is the truth indeed. How could it not be? It is the basis of the *purāṇa*s, the *veda*s, and the *āgama*s.**

The dark *pakṣa* is the part of the lunar month during which the moon is waning. According to an ancient myth, the light of the moon is nectar inside it. When the moon is waning, that is, during the dark *pakṣa*, the nectar is being drunk by the gods, whereas the moon orb is again filled up with nectar when the moon is waxing, that is, during the bright *pakṣa*. Jñānarāja affirms the myth, arguing that it is fundamental in the sacred texts.

(5) **He who is born new and new [again], the leader of the days he goes before the dawns. The moon gives the gods their share, and he prolongs our lives.**

The verse is simply a rendition of *Ṛgveda* 10.85.19.

(6) **In the *veda*, the gods are celebrated as rays of the sun. They extend the digits [of the moon] in a regular fashion during the bright [*pakṣa*], and they remove [them] during the dark [*pakṣa*s]. That is the pronunciation or opinion.**

The rays of the sun, personified here as gods, extend the

digits of the moon when the moon is waxing, and they remove them when it is waning.

(7) **If the two declinations [that is, those of the sun and the moon] have the same direction, take the difference of [the longitudes of] the sun and the moon; otherwise the sum.**

Wherever the moon is with respect to the sun, [its] direction is [given by] the sine of the arc produced from the corrections [?].

The verse aims to determine the distance and direction between the sun and the moon. The first part is clear. It gives the distance, measured along a perpendicular to the ecliptic, between the sun and the moon. The second part, however, is vague.

(8–9^b) **The shadow produced by the moon at noon is multiplied by the [shadow] hypotenuse, and [the result] is subtracted from the sine of the [local] latitude multiplied by 12. If the direction [of the shadow] is to the south, it is added to [rather than subtracted from that quantity]. [The result] divided by the sine of the perpendicular to it [?] is the leg.**

The upright is measured by 12, and the hypotenuse is [achieved] from this, as explained by the best of the sages for the sake of joy.

The quantities found here are the leg, upright, and hypotenuse that will be used in the following verses to contract the diagram. However, the formula given for the leg is unclear. The upright is to have the length 12, and the hypotenuse is to be found in the usual way, that is, by means of the Pythagorean theorem. As will be seen below, the quantities are clear in the diagram; however, they are meant to be scaled by Jñānarāja.

(9^c–10^b) **The arc minutes of [the longitude of] the sun diminished by [the longitude of] the moon are divided by 900; in this case [we get] the bright [portion of the disk of the moon].**

For the sake of the diagram, the [result] is multiplied by the number of *aṅgula*s in the disk of the moon and divided by 12.

When the moon is 180° from the sun, its full disk is illumined. Jñānarāja wants this angular distance in arc minutes, so the disk is fully illumined when there are 10,800′ between the sun and the moon.

Let D be the angular distance between the sun and the moon in arc minutes, and let d be the diameter of the moon in *aṅgula*s. Jñānarāja's expression for the illumined part of the moon to be used for his diagram is

$$\frac{D}{900} \cdot \frac{d}{12} = \frac{D}{10{,}800} \cdot d. \tag{199}$$

This is the usual formula for the illuminated portion of the disk of the moon.

(10[c]) **Having made the mark designated to be the sun on even ground, the leg [is placed] with respect to that in its given direction.**

The given direction is either north or south, depending on whether the moon is north or south of the ecliptic.

(11) **[First] drawing the upright from the tip [of the leg] in the direction [away from the sun] and perpendicular [to the leg], [then] drawing the hypotenuse from the tip of the [upright] to the tip of the leg, the disk of the moon at the given time is there at the tip of the upright. The directions are to be determined [by whether the moon is west or east of the sun].**

The construction is shown in Figure 31. The point A represents the sun, M is the center of the moon, and the circle $NWSE$ is the disk of the moon. The length AB is the leg, BM the upright, and AM the hypotenuse. North, west, south, and east are marked on the moon as indicated by the points N, W, S, and E.

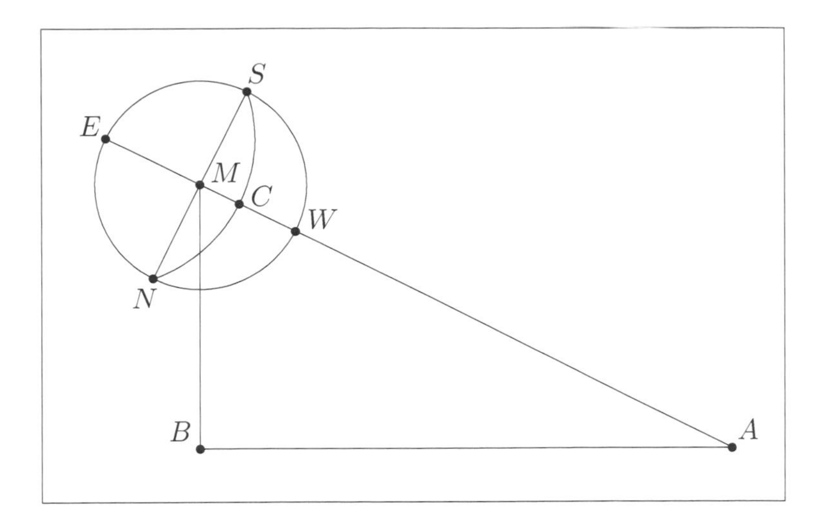

Figure 31: Diagram for the horns of the moon

(12) **The west-east line [on the disk of the moon] is overlapping with the hypotenuse, [extending] from the intersection of the hypotenuse and the disk [of the moon to the point opposite on the disk]. The bright [portion of the moon], [measured] in *aṅgulas* and so on, is [to be marked] inside [of the disk] on the east-west line [on the disk of the moon] extending the hypotenuse up to the intersection of the [line] and the disk [of the moon].**

Two fish figures are [then] to be drawn by means of the distance between the southern and northern parts [?].

The construction of the fish figures is unclear. Two fish figures could have been used to draw a perpendicular to the east-west line on the disk of the moon, but how so is not clear.

The point C in Figure 31 represents the bright portion measured in *aṅgulas*, mentioned in the verse. As such, the bright part of the lunar disk is represented by the figure $SCNW$ and the dark portion by $SCNE$.

(13) **Having drawn a circle from [that is, centered at] the intersection of the pair of strings connecting the heads and tails of the two fish, the obscured part of the**

moon touches [each of] the three points, as the horns of the moon are raised.

The circle constructed is one that passes through the three points, and the construction should give its center. However, this part is unclear, particularly where Jñānarāja would have the center placed.

(14) **During the dark [*pakṣa*], it is opposite. Subtracting [the longitude of] the sun increased by 6 zodiacal signs from [the longitude of] the moon, the bright [portion of the moon] is to be computed as before; the elevation of the moon is visible.**

The words *śambhujaye ca datvā* of the verse are unclear and left untranslated.

The previous construction assumed that the moon was waxing. If the moon is waning, one adds six zodiacal signs, that is, 180° to the longitude of the sun and subtracts it from the longitude of the moon, after which the construction is carried out as before.

(15) **What is said in the *Siddhāntaśiromaṇi*, "When there is no upright [that is, it is of length 0], the support of the elevation of the horns" [and so on], there is no observation of that [statement] on the day of the new moon, at conjunction of the moon and the sun, and on the horizon. How is there an elevation of the horns of the moon where the horizon supports the elevation? How is there nonexistence of the upright?**

Jñānarāja criticizes Bhāskara II's rule, which includes the case where there is no upright.

(16) **The south-north distance between [the position of] the moon and [the position of] the sun is the leg. The sun is at its root. The upright is attached to the tip of the leg, located above it.**

If it is asked, "How is the disk of the moon there at its tip?" When the elevation of the sun is running all

the way to the moon [?], then the inclinations [of the sun and the moon] run south-north.

The first part of the verse defines the leg, upright, and hypotenuse, as has already been used for the diagram. The second part of the verse is unclear.

(17) **The string above the sun attached to the orb of the moon, whatever circle is [described] by means of that [as radius], having the form of half the orb of the moon. To what extent the measure of that is on the surface of the moon, to that extent it is derived from the course of the leg, upright, and hypotenuse, its position in the part of the sky obtained from the sun.**

The meaning of the verse is unclear.

(18) **Thus the section on the elevation of the lunar horns is explained in the beautiful and abundant *tantra* composed by Jñānarāja, the son of Nāganātha, which is the foundation of [any] library.**

16. Chapter on Mathematical Astronomy

Section 10: *Conjunctions of Planets*

This section investigates the occurrence of conjunctions of planets, including computation of the time for a conjunction.

(1–2[b]) **After applying the visibility correction and the *ayanavalana* to the corrected [longitudes of two given] planets based on the given directions, the minutes of arc in the difference [of the longitudes] divided by the difference of the [true] velocities [of two planets] yield [the time of] the conjunction of the planets.**

If one planet is retrograde, [but not the other, the time of conjunction is found] by means of the days resulting from the sum of their [true] velocities.

The visibility correction and the *ayanavalana* have already been discussed; see *Siddhāntasundara* 2.7.7 and 2.5.22–32, respectively. The results are the actual position of the planets in the sky, as seen from a given location on the surface of the earth.

The formula given is straightforward. The difference of the true longitudes of the two planets gives the angular distance to be traversed, while the difference of the true velocities gives the velocity by which the two planets approach (or move away from) each other. Time is then found as distance over velocity.

The formula assumes that the two planets are moving in the same direction, that is, that they either both have direct motion or both have retrograde motion. If one of the planets in question is retrograde and the other is not, the two planets move in opposite directions and it is therefore necessary to use the sum of their velocities for the computation described in the previous verse rather than the difference.

($2^{c–d}$) **When the retrograde planet has the smaller velocity or the smaller [longitude of the two planets], [the conjunction] has already occured; [if it is the] opposite, [the conjunction] will occur in the future.**

If the planet with retrograde motion has the smallest longitude, the two planets are moving away from each other, so that the conjunction has already occurred. However, it is not correct that just because the retrograde planet has a smaller velocity than the planet with direct motion, the conjunction has occurred; the planets could be close and moving toward each other.

(3) **When both planets are retrograde, they advance separately according to the obtained days [from the computation given before]. The latitudes of the planets are the same in this case. The computed latitude of the moon is corrected by the latitudinal parallax.**

The meaning of the verse is unclear. In particular, the latitudes of two retrograde planets are not necessarily equal, and there is no reason to limit a correction by the latitudinal parallax to the moon alone.

($4–5^{b}$) **The distance between [the centers of] two planets [with equal longitudes], which are located in the directions given by their respective celestial latitudes, is [found as] the difference or the sum of their latitudes when they are [respectively] the same or different.**

When the apparent latitude, which is the distance [between the centers of the two planets], is less than half the sum of [the diameters of] the disks [of the planets], then there is a [conjunction known as] *bhedayoga*, just as in the case of a solar eclipse.

A conjunction means that the two planets in question occupy the same position with respect to the ecliptic at the same time. A *bhedayoga* conjunction (sometimes also called *bhedayuti*), however, occurs when the disks of the planets actually overlap in the sky, rather than merely having the same position. As such, a *bhedayoga* is what most people understand by "a conjunction of two planets," namely, that one of the planets obscures the other.

(5^{c}–6) **In the case of Mars, [when it is at its mean distance from the earth,] its disk extends 1,885 *yojanas*. The measure of the disk of Mercury is 289 [*yojanas*]. [That] of Jupiter is 16,602 [*yojanas*]. The disk of Venus is 1,112 [*yojanas*]. [That] of Saturn is 29,646 [*yojanas*].**

The diameters in *yojanas* of the disks of the five star-planets at their mean distances from the earth are given; these are called mean diameters. The results are summarized in Table 18.

Jñānarāja has already, in *Siddhāntasundara* 2.5.7^{a-b}, given the mean diameters of the sun and the moon.

(7) **[The diameters in *yojanas* of the disks] are multiplied by the difference of [the given planet's] mean distance from the center of the earth and [its] actual distance, and [then] divided by three times the sine of the greatest equation. [The diameters in *yojanas* of the disks] diminished or increased by the result, when the geocentric distance is greater or less than the radius R [respectively], are thus the corrected diameters of the disks of the planets.**

The formula given is unusual and not entirely clear. What Jñānarāja finds by subtracting the mean geocentric distance from the true geocentric distance is the distance of the planet from its mean orbit. When that distance is as large as it can be, it is correlated to three times the maximum equation, though the exact relationship is unclear. Compare *Siddhāntasundara* 2.4.4.

(8) **The disks of the planets measured in minutes of arc and beginning with Mars are 5 [Mars], 6 [Mercury], 7[Jupiter], 9 [Venus], and 5 [Saturn].**

Having, as previously, determined the apparent [diameters], dividing them by 3 is [the disk measured in] *aṅgulas*; this is due to the remoteness [of the planets].

Planet	Diameter, *yojanas*	Diameter, arc minutes
Mars	1,885	5
Mercury	289	6
Jupiter	16,602	7
Venus	1,112	9
Saturn	29,646	5

Table 18: Mean diameters of the planets

This verse gives the mean diameters in arc minutes. The information is given in Table 18.

Since there are 3 *aṅgula*s per arc minute, dividing the diameter in arc minutes by 3 converts it to *aṅgula*s.

(9) **Thus is the conjunction of planets accomplished in the beautiful and abundant *tantra* composed by Jñānarāja, the son of Nāganātha, which is the foundation of [any] library.**

17. Chapter on Mathematical Astronomy

Section 11: *Occurrence of Pātas*

The Sanskrit term *pāta*, in addition to denoting the node of a planet, also denotes peculiar configurations of the sun and the moon, as will be described below. The interest in finding when these configurations occur is due to their ominous nature.

(1) **The ancients say: "If the sun and the moon are on the same or different [side] of the equator when the equinox is derived from the sum [of the true longitudes] of the moon and the sun [?], then the *pāta* [that is, the inauspicious moment] causes a downfall, like during a solar eclipse."**

The reading of the verse is difficult, but it clearly defines a *pāta* as an inauspicious time when the positions of the sun and the moon meet certain requirements. In the Indian tradition, a *pāta* occurs when the sun and the moon have the same declination and are at the same or opposite sides of the equator, though the present verse does not appear to address the declinations of the two luminaries.

The meaning of a "downfall" is not specified, but is probably intended to signify a generally inauspicious time.

(2) **On the day near *brahmadhruva* [?], when the sum [of the true longitudes] of the sun and the moon is a circle [that is, 360°] or half a circle [that is, 180°], the *pāta* is, respectively, called *vaidhṛta* and *vyatīpāta*.**

The word *pāta* is used as a common term to designate both *vyatīpāta* and *vaidhṛta*. If the sum of the longitudes of the sun and the moon equals 360° and the two luminaries have the same magnitude in declination but in different directions, we have a *pāta* called *vaidhṛta*. However, if the sum of the longitudes equals 180° and the two luminaries have the same magnitude in declination in the same direction, then the *pāta* is known as *vyatīpāta*.

What is meant by *brahmadhruva* is unclear.

(3) **The minutes of arc of the *pāta* [that is, of the sum of the true longitudes of the sun and the moon] diminished or increased by [360° or 180°, as the case may be] are divided by the sum of the [true] velocities [of the sun and the moon]; the result is the days to elapse or that have lapsed [until or since the *pāta*]. The *pāta*s of the sun and the moon occur by means of the [days thus computed]; [first] the lapsed *pāta*, [then] afterward the future [one]; the *pāta*s of the sun, the moon, and the moon [occur] at those times.**

While the formulation used here is somewhat clumsy—note, for example, the repetition of "the moon" in the last sentence—the formula presented is clear: the sum of the true longitudes of the sun and the moon is subtracted from 21,600′ (that is, 360°) or 10,800′ (that is, 180°), depending on whether we are interested in, respectively, *vaidhṛta* or *vyatīpāta*. The result is the "angular distance" to the *pāta*, which either has occurred or will occur at a future time. The time is then found as distance over velocity, where the sum of the velocities of the sun and the moon is used since the two luminaries move in the same direction.

If we assume, for example, that we are interested in *vyatīpāta*, the expression for the time becomes

$$\frac{10{,}800' - (\lambda_{\odot} + \lambda_{☾})}{v_{\odot} + v_{☾}}.$$

The formula is similar if we are dealing with *vaidhṛta*.

(4) **When [the sum of the true longitudes] is 0°, 180°, or 360°, there is equality of the declinations [of the sun and the moon] when the latitude [of the moon] is great.**

When the moon is in an odd or an even quadrant, that is, in the same or different hemisphere than [the

longitude of] the moon diminished by [the longitude of] the [lunar] node, [then] [?].

The contents of the verse are not entirely clear, other than it speaks of the equality of the declinations of the sun and the moon at the time of a *pāta*, though no mention is made of the directions of them. The last sentence appears incomplete and does not seem to carry naturally into the next verse.

(5) **The declination of the sun is to be computed, and likewise the latitude of the moon; then the method of *bījas* is to be applied.**

A set of *bījas* is often a means to convert data from one astronomical school to another, or to make corrections to data (see commentary on *Siddhāntasundara* 2.1.83–84). It follows that Jñānarāja wants to correct the computed declination of the sun and latitude of the moon with *bījas*, but no further details are given, and thus it is unclear what *bījas* he has in mind here, or how they are to be applied.

(6) **By however much the amount of the mean declination of the moon is measured, it is to be corrected by the minutes of arc in its latitude; [the result] is the true declination of the moon. If the true declination of the moon is equal to the declination of the sun, making the two *pakṣas* equal [?], the result is the measure of the mean declination of the moon.**

Since the moon has a latitude, its true declination is found from its mean declination corrected for the current latitude. The meaning of the rest of the verse is unclear. The two *pakṣas* referred to are the two halves, the bright and the dark, of the lunar month.

(7–8) **One should then compute the arc of the declination by means of *koṣṭhakas*, [giving] the degrees of the arc. The difference between the arcs produced by the moon at a given time and at a previous time divided by the [true] velocity [of the moon gives the time it took**

the moon to go from one declination to the other].

As before, the declination is to be employed by the obtained days measured by the result, elapsed by the future ones [?], by the counted-up [?] parts measured by 90 in the celebrated *koṣṭhaka*s. Also, one should again, as before, compute the days lapsed or to lapse by half of that, by the *pāta*, the sun, and the moon; equality of the declinations is exceedingly pleasing.

For *koṣṭhaka*s, see the discussion of astronomical treatises in the Indian tradition (Introduction, page 19). It appears that Jñānarāja is suggesting that tables are a good tool to have handy when working out the time of a *pāta*.

The last half of verse 7 and verse 8 make little sense. The topic is equality of the declinations of the sun and the moon, but the passage appears to be corrupt.

(9) **Since the declination of the moon is increasing [when it is] greater than the declination of the sun, or else it is decreasing [when it is] less [than the declination of the sun], therefore the future *pāta* is reversed.**

The argument presented is hard to follow. For example, it does not seem to be correct that the declination of the moon is increasing when it is greater than that of the sun, unless some constraints are introduced.

(10–11^b) **The degrees of the declination [of the sun] are given [as follows] in intervals of** $10°$ **[of the sun's distance to an equinoctial point]:** $4°3'$, $8°$, $11°45'$, $15°9'$, $18°10'$, $20°37'$, $22°28'$, $23°37'$, **and** $24°$.

Given here is a table of declinations of the sun, summarized in Table 19. The arc is the distance of the sun from an equinoctial point; the corresponding declination is given.

(11^c–12^b) **The arc minutes of the latitude [of the moon] are in steps of** $10°$ **[of the moon's distance from its node]:** $47°$, $92°$, $135°$, $173°$, $206°$, $234°$, $253°$, $266°$, **and** $270°$.

Arc	Declination of the sun
$10°$	$4°3'$
$20°$	$8°$
$30°$	$11°45'$
$40°$	$15°9'$
$50°$	$18°10'$
$60°$	$20°37'$
$70°$	$22°28'$
$80°$	$23°37'$
$90°$	$24°$

Table 19: Declinations of the sun

Arc	Lunar latitude
$10°$	$47'$
$20°$	$92'$
$30°$	$135'$
$40°$	$173'$
$50°$	$206'$
$60°$	$234'$
$70°$	$253'$
$80°$	$266'$
$90°$	$270'$

Table 20: Latitudes of the moon

Given here is a table of lunar latitudes, summarized in Table 20. The arc is the distance of the moon from a node; the corresponding lunar latitude is given.

If the latitude is needed at a point where the moon's distance to its node is not multiple of 10°, linear interpolation is used; this is the same approach as was used for the Sine tables in the section on true motion. For example, if the moon is 32° from its node, its declination δ is

$$\delta = 135' + (32° - 30°) \cdot \frac{173' - 135'}{40° - 30°} = 142'36''.$$

(12^{c-d}) **The sine of the declination and its direction is**

[derived] from [the longitude of] the precession-corrected planet, and the [lunar] latitude is [derived] from [that] sine by means of an interval [?] of 90°.

The first statement made here is, of course, true; the declination of any planet must be derived from the precession-corrected longitude of that planet. However, the statement regarding the derivation of the (in this case) lunar latitude from the declination is less clear. In fact, the magnitude of the latitude of a planet depends on the planet's angular distance to its nodes, not on the declination.

(13) **[If] the equality of the declinations, west and east [?], is from the time of equality of the declinations, then it should be west [?].**

The declination of the moon at that time grows or decreases if diminished or increased by its latitude.

The first statement is again unclear, whereas the last statement of the verse is obvious, unless, of course, the lunar latitude is $0'$.

(14) **The time of the *madhyapāta* is [when] the declinations of the centers of the disks of the sun and the moon are equal.**

Up to the difference [between the declinations] is less than half the sum [of the diameters] of the disks [of the sun and the moon], there is equality of the declinations on the horizon.

The definition of *madhyapāta* is the time when the declinations of the centers of the sun and the moon are equal. It is the "mid-time" of the *pāta*. The last part of the verse makes less sense. If the difference of the declinations of the sun and the moon is less than half the sum of the diameters of the disks of the sun and the moon, there is still an "overlap," so to speak, but the temporal aspect is unclear.

(15) **From the mean declination of the moon diminished by the arc minutes in half the sum of the measures [of**

the disks] the moon is diminished by its moon [?].

As in the case of an eclipse, the duration [of the *pāta*] is divided by the degrees and arc minutes of the velocity are the *ghaṭikā*s [that the *pāta* lasts], in the western and eastern part [?].

The first part of the verse is difficult and makes little sense as it stands. The second part summarizes a fairly clear point, namely, that if the duration of a *pāta* is divided by the velocity involved (in this case, the sum of the velocities of the sun and the moon), the total time of the *pāta* is the result. Compare *Siddhāntasundara* 2.5.15–16.

(16) **At whatever degree[s] on [their] own arc[s] the slowly changing declination of the sun or of the true moon is, the difference between them multiplied by 60 and divided by the [true] velocity of the moon. By the lapsed or future days given by the result, which bring to motion the *pāta*s of the sun and the moon, there is again equality of the declinations [of the sun and the moon]. In this manner it is easily computed for those of strong intellects.**

The formula appears to aim at computing the next time, either in the past or upcoming, when the declinations of the sun and the moon are equal. However, as it stands, the formula cannot be correct. For one, the declinations of the sun and the moon both play a role, but only the velocity of the moon is used.

(17) **Thus is the section on *pāta*s, accompanied by its method, explained in the beautiful and abundant *tantra* composed by Jñānarāja, the son of Nāganātha, which is the foundation of [any] library.**

References

Aufrecht, T. (1962). *Catalogus Catalogorum: An Alphabetical Register of Sanskrit Works and Authors.* Wiesbaden: Franz Steiner Verlag. Two volumes. Part I is in volume 1; parts II and III in volume 2. Reprint of the 1891 (part I) and 1896 (parts II) and 1903 (part III) edition published by F. A. Brockhaus, Leipzig.

Bag, A. K. (1979). *Mathematics in Ancient and Medieval India*, Volume 16 of *Chaukhambha Oriental Research Studies.* Varanasi and Delhi: Chaukhambha Orientalia.

Bendall, C. (1902). *Catalogue of the Sanskrit Manuscripts in the British Museum.* London: British Museum.

Benson, J. (2001). Śaṃkarabhaṭṭa's Family Chronicle: The Gādhivaṃśavarṇana. In A. Michaels (ed.), *The Pandit: Traditional Scholarship in India*, pp. 105–118. New Delhi: Manohar Publications.

Berggren, J. L. (2002). Some Ancient and Medieval Approximations to Irrational Numbers and Their Transmission. In Y. Dold-Samplonius, J. W. Dauben, M. Folkerts, and B. Van Dalen (eds.), *From China to Paris: 2000 Years Transmission of Mathematical Ideas*, pp. 31–44. Stuttgart: Franz Steiner Verlag.

Bhatt, G. H., et al. (1960–75). *The Vālmīki Rāmāyaṇa: Critical Edition.* Baroda: Oriental Institute. 7 volumes.

Bronner, Y. (2010). *Extreme Poetry: The South Asian Movement of Simultaneous Narration.* New York: Columbia University Press.

Bühnemann, G. (1991). Selecting and Perfecting Mantras in Hindu Tantrism. *Bulletin of the School of Oriental and African Studies, University of London 54*(2), 292–306.

Bühnemann, G. (1992). On Puraścaraṇa: Kulārṇavatantra, Chapter 15. In T. Goudriaan (ed.), *Ritual and Speculation in Early Tantrism: Studies in Honor of André Padoux*, pp. 61–106. Albany: State University of New York Press.

Burgess, E. (1858–60). Translation of the Sûrya-Siddhânta, A Text-Book of Hindu Astronomy; With Notes, and an Appendix. *Journal of the American Oriental Society 6*, 141–498.

Chakrabarti, G. (1934). Surd in Hindu Mathematics. *Journal of the Department of Letters (Calcutta University) 24*, 29–58.

Chatterjee, B. (ed.) (1981). *Śiṣyadhīvṛddhida Tantra of Lalla with the Commentary of Mallikārjuna Sūri.* New Delhi: Indian National Science Academy. Two volumes.

Colebrooke, H. T. (1977). *Essays on History, Literature and Religions of India (Miscellaneous Essays).* New Delhi: Cosmo Publications. Reprint, first published under the title "Miscellaneous Essays" in 1837.

Datta, B. (1927). On Mûla, the Hindu Term for "Root." *American Mathematical Monthly 34*(8), 420–423.

Datta, B., and A. N. Singh (1993a). Approximate Values of Surds in Hindu Mathematics. *Indian Journal of History of Science 28*(3), 265–275. Revised by Kripa Shankar Shukla.

Datta, B., and A. N. Singh (1993b). Surds in Hindu Mathematics. *Indian Journal of History of Science 28*(3), 253–264. Revised by Kripa Shankar Shukla.

Dhavale, D. G. (1969). Saptarṣicāra. *Journal of the University of Poona, Humanities Section 29*, 69–81.

Dhavale, D. G. (ed.) (1996). *The Brahmasiddhānta of Śākalyasaṃhitā, Critically Edited with Introduction and Appendices.* Pune: Bhandarkar Oriental Research Institute.

Dikshit, S. B. (1969–81). *English Translation of Bharatiya Jyotish Sastra (History of Indian Astronomy).* Delhi: Civil Lines. Translated by R. V. Vaidya.

Duke, D. (2005). The Equant in India: The Mathematical Basis of Ancient Indian Planetary Models. *Archive for His-*

tory of Exact Sciences 59, 563–576.

Dvivedi, V. P. (ed.) (1912). *Jyautisha Siddhánta Sangraha: Ancient Hindu Astronomical Works: Somasiddhánta & Brahmasiddhánta*, Volume 38 of *Benares Sanskrit Series*. Benares: Braj Bhushan Das & Co.

Dvivedī, V. P. Ś. (ed.) (1907). *Vasiṣṭhasiddhānta*. Banārasa: Vidyāvilāsa Presa (Press). This is the same editor as the editor of the Jyautisha Siddhánta Sangraha.

Dvivedin, S. (ed.) (1902). *Brāhmasphuṭasiddhānta and Dhyānagrahopadeśādhyāya by Brahmagupta*. Benares: Medical Hall. Reprint from the Pandit. Edited with editor's own commentary.

Eggeling, J., et al. (1886–94). *Catalogue of the Sanskrit Manuscripts in the Library of the India Office*, Volume 1, Part 5. London: Printed by the order of the Secretary of State for India in Council. (Volume 1 contains parts 1–7 and Volume 2 parts 8–9.).

Evans, J. (1998). *The History & Practice of Ancient Astronomy*. New York and Oxford: Oxford University Press.

Firishtah, M. Q. H. S. A. (1971). *History of the Rise of the Mahomedan Power in India, Till the Year A.D. 1612, Translated from the Original Persian of Mahomed Kasim Ferishta by John Briggs, M.R.A.S., Lieutenant-Colonel in the Madras Army, To which is Added, An Account of the Conquest by the Kings of Hydrabad, of those Parts of the Madras Provinces Denominated, The Ceded Districts and Northern Circars, With Copious Notes, In Four Volumes.* Calcutta: Editions India. Reprint of first edition, London, 1829.

Grant, E. (1987). Celestial Orbs in the Latin Middle Ages. *Isis 78*(2), 152–173.

Haig, W. (ed.) (1928). *The Cambridge History of India*, Volume 3. Cambridge: Cambridge University Press.

Ikeyama, S. (2002, May). *The Brāhmasphuṭasiddhānta Chapter 21 with the Commentary of Pṛthūdakasvāmin*. PhD thesis, Brown University.

Jain, P. K. (ed.) (2001). *The Sūryaprakāśa of Sūryadāsa (a Commentary on Bhāskarācārya's Bījagaṇita)*, Volume

182 of *Gaekwad's Oriental Series*. Vadodara: Oriental Institute. Volume I: A Critical Edition English Translation and Commentary for the Chapters Upodghāta, Ṣaḍvidhaprakaraṇa and Kuṭṭakādhikāra.

Jinavijaya, P. M. (ed.) (1965). *A Catalogue of the Sanskrit and Prakrit Manuscripts in the Rajasthan Oriental Research Institute (Jodhpur Collection), Part II (B)*, Volume 81 of *Rajasthan Puratana Granthamala*. Jodhpur: Rajasthan Oriental Research Institute.

Knudsen, T. L. (2008, May). *The Siddhāntasundara of Jñānarāja: A Critical Edition of Select Chapters with English Translation and Commentary*. PhD thesis, Brown University.

Knudsen, T. L. (2009). Crystalline Spheres in the Siddhāntasundara of Jñānarāja. In G. Gnoli and A. Panaino (eds.), *Kayd: Studies in History of Mathematics, Astronomy and Astrology in Memory of David Pingree*, pp. 55–58. Roma: Istituto Italiano per l'Africa et l'Oriente.

Lallurama, S. J., M. G. Bakre,, and D. V. Gokhale (eds.) (2001). *The Bhagavad-gītā with Eight Commentaries: (1) Tattvaprakāśikā of Keśavakaśmīri Bhaṭṭācārya; (2) Guḍhārthadīpikā of Madhusūdana Sarasvatī; (3) Tātparyabodhinī of Śaṅkarānanda; (4) Subodhinī of Śrīdharasvāmin; (5) Bhāvaprakāśa of Sadānanda; (6) Bhāṣyotkarṣadīpikā of Dhanapatisūri; (7) Paramārthaprapā of Daivajña Paṇḍita Sūrya; and (8) Arthasaṅgraha of Rāghavendra*, Volume 59 of *Parimal Sanskrit Series (parimala saṃskṛta granthamālā)*. Delhi: Parimal Publications. Three volumes. Reprint of the Gujarati Printing Press edition, Bombay 1915.

Lalye, P. G. (ed.) (1979). *Navarasamañjarī and Sūryodayakāvya: Two Sanskrit Manuscripts*. Hyderabad: P. G. Lalye.

Minkowski, C. (2002). Astronomers and Their Reasons: Working Paper on Jyotiḥśāstra. *Journal of Indian Philosophy 30*, 495–514.

Minkowski, C. (2004a). Competing Cosmologies in Early

Modern Indian Astronomy. In C. Burnett, J. P. Hogendijk, K. Plofker, and M. Yano (eds.), *Studies in the History of the Exact Sciences in Honour of David Pingree*, pp. 349–385. Leiden and Boston: Brill.

Minkowski, C. (2004b). On Sūryadāsa and the Invention of Bidirectional Poetry (vilomakāvya). *Journal of the American Oriental Society 124*(2), 325–333.

Minkowski, C. (2011). Seasonal Poetry as Science: The ṛtuvarṇana in Some Astronomy Treatises. *Gaṇita Bhāratī 33*(1–2), 89–113.

Minkowski, C., and T. Knudsen (2011). The ṛtuvarṇana Chapter of Jñānarāja's Siddhāntasundara: Text, Translation, and Notes. *Gaṇita Bhāratī 33*(1–2), 55–87.

Navathe, P. D. (1991). *Descriptive Catalogue of Manuscripts in the Government Manuscripts Library Deposited at the Bhandarkar Oriental Institute, Volume III, Part IV.* Poona: Bhandarkar Oriental Research Insitute.

Ohashi, Y. (1994). Astronomical Instruments in Classical Siddhāntas. *Indian Journal of History of Science 29*(2), 155–313.

Patte, F. (ed.) (2004). *L'œuvre mathématique et astronomique de Bhāskarācārya: Le Siddhāntaśiromaṇi.* Genève: Droz. Two volumes.

Peterson, P. (1892). *Catalogue of the Sanskrit Manuscripts in the Library of His Highness the Maharaja of Ulwar.* Bombay.

Pingree, D. (1970). Āryabhaṭa. In C. C. Gillespie (ed.), *Dictionary of Scientific Biography*, Volume 1, pp. 308–309. New York: Charles Scribner's Sons.

Pingree, D. (1970–94). *Census of the Exact Sciences in Sanskrit.* Philadelphia: American Philosophical Society. Series A, Volumes 1–5.

Pingree, D. (1971). On the Greek Origin of the Indian Planetary Model Employing a Double Epicycle. *Journal for the History of Astronomy 2*(2), 80–85.

Pingree, D. (1972). Precession and Trepidation in Indian Astronomy before A.D. 1200. *Journal for the History of As-*

tronomy 3, 27–35.

Pingree, D. (1973). Keśava. In C. C. Gillespie (ed.), *Dictionary of Scientific Biography*, Volume 7, pp. 314–316. New York: Charles Scribner's Sons.

Pingree, D. (1976a). Varāhamihira. In C. C. Gillespie (ed.), *Dictionary of Scientific Biography*, Volume 13, pp. 581–583. New York: Charles Scribner's Sons.

Pingree, D. (ed.) (1976b). *Vṛddhayavanajātaka of Mīnarāja*, Volumes 162 and 163 of *Gaekwad's Oriental Series*. Baroda: Oriental Institute.

Pingree, D. (1978a). History of Mathematical Astronomy in India. In C. C. Gillespie (ed.), *Dictionary of Scientific Biography*, Volume 15, pp. 533–633. New York: Charles Scribner's Sons.

Pingree, D. (1978b). Islamic Astronomy in Sanskrit. *Journal for the History of Arabic Science 2*(2), 315–330.

Pingree, D. (1981). *Jyotiḥśāstra: Astral and Mathematical Literature*, Volume 6 of *A History of Indian Literature*. Wiesbaden: Otto Harrassowitz. Volume VI, Fasc. 4.

Pingree, D. (1984). *A Descriptive Catalogue of the Sanskrit and Other Indian Manuscripts of the Chandra Shum Shere Collection in the Bodleian Library, Part I, Jyotiḥśāstra*. Oxford: Clarendon Press.

Pingree, D. (1990). The Purāṇas and Jyotiḥśāstra: Astronomy. *Journal of the American Oriental Society 110*(2), 274–280.

Pingree, D. (1997). *From Astral Omens to Astrology: From Babylon to Bīkāner*. Rome: Istituto italiano per l'Africa et l'Oriente.

Plofker, K., and T. L. Knudsen (2008a). Āryabhaṭa. In P. T. Keyser and G. L. Irby-Massie (eds.), *Encyclopedia of Ancient Natural Scientists: The Greek Tradition and Its Many Heirs*. London and New York: Routledge.

Plofker, K., and T. L. Knudsen (2008b). Varāhamihira. In P. T. Keyser and G. L. Irby-Massie (eds.), *Encyclopedia of Ancient Natural Scientists: The Greek Tradition and Its Many Heirs*. London and New York: Routledge.

Santhanam, R. (1984). *Brihat Parasara Hora Sastra of Maharshi Parasara.* New Delhi: Ranjan Publications. Two volumes. Sanskrit text with English translation and commentary.

Sarma, K. M. K. (1950). Siddhānta-saṃhitā-sāra-samuccaya of Sūrya Paṇḍita. *Siddha Bhāratī 2*, 222–225.

Sarma, K. V. (2003). Numerical and Alphabetic Numerical Systems in India. In A. K. Bag and S. R. Sarma (eds.), *The Concept of Śūnya*, pp. 37–71. New Delhi: Indira Gandhi National Centre for the Arts, Indian National Science Academy, and Aryan Books International.

Sarma, S. R. (2008, September). Sine Quadrant in India: Sanskrit Text and Extant Specimens. Paper read at the 14th World Sanskrit Conference in Kyoto, Japan.

Sarma, S. R. (2011). Shabnumā-wa-Rūznumā: A Rare Astronomical Instrument Extant in Two Specimens. *Tarikh-e Elm (Iranian Journal for the History of Science) 9*, 21–48.

Sarma, S. R. (2012). The Dhruvabhrama-Yantra of Padmanābha. *Saṃskṛtavimarśaḥ (Journal of the Rashtriya Sanskrit Sansthan) 6*, 321–343.

Śāstrī, G. D. (ed.) (1999). *The Siddhānta Śiromani: A Treatise on Astronomy by Bhāskarācārya with His Own Exposition The Vāsanābhāshya* (third ed.), Volume 72 of *The Kashi Sanskrit Series.* Chaukhambha Sanskrit Sansthan. Revision of the edition by Bāpū Deva Śāstrī.

Scharf, P. (2008). Sāṃkhya. In D. Cush, C. Robinson, and M. York (eds.), *Encyclopedia of Hinduism*, pp. 745–747. London and New York: Routledge.

Schreiner, P., and R. Söhnen (eds.) (1987). *Sanskrit Indices and Text of the Brahmapurāṇa.* Wiesbaden: Otto Harrassowitz.

Sen, S. N. (1966). *A Bibliography of Sanskrit Works on Astronomy and Mathematics.* New Delhi: National Institute of Sciences in India.

Sharma, O., and B. K. Singh (eds.) (1984). *Catalogue of Sanskrit and Prakrit Manuscripts (Jodhpur Collection), Part*

XVI, Volume 148 of *Rajasthan Puratana Granthamala*. Jodhpur: Rajasthan Oriental Research Institute.

Shastri, J. L. (ed.) (1988). *Bhāgavata Purāṇa of Kṛṣṇa Dvaipāyana Vyāsa with Sanskrit Commentary Bhāvārthabodhinī of Śrīdharasvāmin*. Delhi: Motilal Banarsidass. Reprint of 1983 edition.

Shukla, K. S. (1986). *Vaṭeśvara-siddhānta and Gola*. New Delhi: Indian National Science Academy. Two volumes. Volume 1: Sanskrit text. Volume 2: English translation and commentary.

Smith, D. E. (1951–53). *History of Mathematics*. Boston, New York, etc.: Ginn and Company. Two volumes. Reprint of the 1923–25 edition.

Smyly, J. G. (1944). Square Roots in Heron of Alexandria. *Hermathena 63*, 18–26.

Staff of the Manuscripts Section (1963). *A Descriptive Catalogue of the Sanskrit Manuscripts, Acquired for and Deposited in the Sanskrit University Library (Sarasvati Bhavana), Varanasi, During the Years 1791–1950, Volume IX, Jyautiṣa Mss*. Varanasi: Sanskrit University. Volume IX, Jyautiṣa Mss.

Śukla, S. (ed.) (2008). *Siddhāntasundaram of Śrī Jñānarāja*, Volume 65 of *Sarasvatībhavana-granthamālā*. Vārāṇasī: Sampūrṇānanda Saṃskṛta Viśvavidyālaya.

Sukthankar, V. S., et al. (eds.) (1933–66). *The Mahābhārata, for the First Time Critically Edited*. Pune: Bhandarkar Oriental Research Institute. 19 volumes.

Tannery, P. (1886). Notice sur les deux lettres arithmétiques de Nicolas Rhabdas (Texte grec et traduction). *Notices et extraits des manuscrits de la Bibliothèque Nationale 32*(1), 121–252.

Trivedi, N., and O. P. Sharma (eds.) (1992). *Catalogue of Sanskrit and Prakrit Manuscripts (Kota Collection), Part XXV*, Volume 169 of *Rajasthan Puratana Granthamala*. Jodhpur: Rajasthan Oriental Research Institute.

Vámanácházrya (ed.) (1882). *Jātaka-paddhati of Keśava Daivajña with a Commentary of Divākara Daivajña*. Kāśī: Meḍikal Hāl (Medical Hall).

Velankar, H. D. (1998). *A Descriptive Catalogue of Sanskrit and Prakrit Manuscripts in the Collection of the Asiatic Society of Bombay, Volume I: Technical Literature* (second ed.). Mumbai: Asiatic Society of Bombay. Compiled by H. D. Velankar. Second edition edited by V. M. Kulkarni and Devangana Desai. In six parts.

Verma, O. P. (1970). *The Yādavas and Their Times.* Nagpur: Vidarbha Samshodhan Mandal.

Weber, A. (1853). *Verzeichniss der Sanskrit-handschriften von Herrn Dr. Weber*, Volume 1 of *Die handschriften-verzeichnisse der Königlichen bibliothek.* Berlin: Nicolai'schen buchhandlung.

Williams, C. J. (2005, May). *Eclipse Theory in the Ancient World.* PhD thesis, Brown University.

Wujastyk, D. (2004). Jambudvīpa: Apples or Plums? In C. Burnett, J. P. Hogendijk, K. Plofker, and M. Yano (eds.), *Studies in the History of the Exact Sciences in Honour of David Pingree*, pp. 287–301. Leiden and Boston: Brill.

Yano, M. (1977). Three Types of Hindu Sine Tables. *Indian Journal of History of Science 12*(2), 83–89.

Youschkevitch, A. P. (1976). *Les mathématiques arabes (VIIIe-XVe siècles).* Paris: Librairie Philosophique J. Vrin. Translated from Russian to French by M. Cazenave and K. Jaouiche.

Index